Russian Planetary Exploration
History, Development, Legacy, Prospects

Brian Harvey

Russian Planetary Exploration

History, Development, Legacy, Prospects

Springer

Published in association with
Praxis Publishing
Chichester, UK

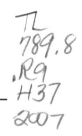
PRAXIS

Brian Harvey
2 Rathdown Crescent
Terenure
Dublin 6W
Ireland

SPRINGER–PRAXIS BOOKS IN SPACE EXPLORATION
SUBJECT *ADVISORY EDITOR*: John Mason, M.Sc., B.Sc., Ph.D.

ISBN 10: 0-387-46343-7 Springer Berlin Heidelberg New York
ISBN 13: 978-0-387-46343-8 Springer Berlin Heidelberg New York

Springer is part of Springer-Science + Business Media (springer.com)

Library of Congress Control Number: 2006938306

Cover design: Jim Wilkie
Project management: Originator Publishing Services, Gt Yarmouth, Norfolk, UK

Printed on acid-free paper

Contents

Author's preface

The many, great and deserved achievements of the United States and Europe in planetary exploration in recent years can obscure the fact that it was the Soviet Union that forged the way to the planets. The idea of flying to Venus and Mars dates to tsarist times and the popular film made in the 1920s about a Mars colony, *Aelita*. Plans for an expedition to Mars were put on the drawing board in Stalin's Soviet Union. It can be forgotten that the Soviet manned moon rocket, the N-1, was originally designed with a view to sending cosmonauts to Mars – until the moon race diverted planners from their original purpose. Calculations for the first unmanned flights to the planets were done within months of the first Sputnik going into orbit.

The Soviet Union achieved many important firsts in its programme of planetary exploration.

- First probes to Venus (1961) and Mars (1962).
- First probe to reach the surface of another world, Venus (1966).
- First soft-landing on Venus (1970).
- First soft-landing on Mars (1971).
- First picture from the surface of Mars (1971).
- First surface analysis of the rocks of another planet, Venus (1972).
- First pictures from the surface of Venus (1975).
- First spacecraft to orbit Venus (1975).
- First on-site laboratory analysis of the rocks and soil of Venus (1982).
- First radar maps of another planet, Venus (1983).
- First balloon to fly in the atmosphere of another planet (1985).
- First close flyby of a comet, Comet Halley (1986).
- First interception of a moon in orbit around another planet, Phobos (1989).

These spaceships were equipped with a sophisticated range of instruments and experiments. The scientific haul from these missions was considerable and transformed our views of the planets Venus and Mars. Results from these missions were made available to the scientific community. Although the Western image of Soviet Mars probes is that they all failed, in reality they contributed much to our early knowledge of the planet.

To this day, four Russian spacecraft still circle Venus and another four orbit Mars. Ten landers still sit on the surface of Venus while three found their way to the surface of Mars. The Soviet Union/Russia launched 54 interplanetary spacecraft, so they were an important part of its programme of space exploration.

Now, 50 years after the first Sputnik, the Russians are making fresh plans to return to the planets. Due to the extreme financial crisis which beset the Russian Federation in its early days, only one deep space probe was launched in the post-Soviet period (Mars 96). The Russian space programme for 2006–2015 includes the recovery of a soil sample mission from the Martian moon Phobos over 2009–2011. Now is a good time to put the Soviet and Russian deep space programme into its historical perspective. It represents an archive of experience and knowledge useful to those planning new missions to Venus, Mars and farther afield. This book tells the story of the Soviet deep space programme: the space probes and their often inventive designs; the people who built them; the key decisions that were made; the scientific outcomes; the knowledge learned; the heartbreaking failures; and also the joyous successes.

Brian Harvey
Dublin, Ireland, 2007

Acknowledgements

The author wishes to acknowledge and thank all those whose assistance made this book possible. Especially he would like to thank: Rex Hall, for advice, information and making available his collection; Phil Clark, for his technical knowledge; Andrew Ball, an expert on planetary landings, for his comments; David Williams, for assistance in the NASA information site on Soviet deep space exploration; Ruslan Kuzmin, for providing insights into Russian planetary exploration and likewise Igor Mitrofanov; Paolo Ulivi, for sharing his own knowledge of Soviet deep space exploration; Andy Salmon, for giving me access to his collection; Larry Klaes, for forwarding technical documents to me; Suszann Parry, for making available information sources and photographs in the British Interplanetary Society; Prof. Evert Meurs, director and Carol Woods, librarian of Dunsink Observatory; and of course Clive Horwood for his support for this project. I am grateful to them all.

Many of the photographs published here come from the author's collection. I would like to thank the many people who generously provided or gave permission for the use of photographs, especially the following:

- Don P. Mitchell, for permission to use his images of Venus and Mars exploration;
- Ted Stryk, for permission to use his enhancements of Venera and Mars images;
- Dominic Phelan, for his images of IMBP: the *bochka* and Anatoli Grigoriev;
- Andy Salmon, for permission to use his images of Mars 3, Mars 96, Mars 98 and Marsokhod;
- Nick Johnson, for permission to use his images of Mars 94 and the Tsiolkovsky probe, taken from *The Soviet Year in Space 1990* by Teledyne Brown Engineering;
- Rex Hall, for his photographs of Venera 8, the VEGA gondola, UMVL; and NASA, for its collection on Soviet space science.

Brian Harvey
Dublin, Ireland, 2007

Figures

MAPS

TABLES

Abbreviations and acronyms

1MV	Number 1 series for Mars and Venus (1M for Mars, 1V for Venus)
AIS	Automatic Interplanetary Station
ANGSTREM	X-ray spectrometer
APS	Autonomous Propulsion System
APX	Alpha particle, proton and X-ray spectrometer
ASPERA	Mass spectrometer and particle imager
AVD	*Avarinoye Vyklyuchennie Dvigatelei* (emergency cut-off of engines)
BOZ	*Blok Obespecheyna Zapushka* (ignition insurance system)
CBPS	Combined Braking and Propulsion System
CNES	Centre National d'Etudes Spatiales
DPI	Accelerometer
DYMIO	Ion spectrometer
DZhVS	Long-duration Venus station
ERTA	*Elektro Raketny Transportniy Apparat*
ESA	European Space Agency
FGB	*Funksionali Gruzovoi Blok* (functional cargo block)
FONEMA	Ion and high-energy spectrometer
GDL	Gas Dynamic Laboratory (in Leningrad)
GDR	German Democratic Republic
GIRD	Group for the Study of Reactive Propulsion/Devices
glasnost	Openness
GLONASS	Navigation systems
GPS	Global Positioning System
GSMZ	*Imeni Semyon Lavochkin* (state union machine building plant dedicated to the memory of Semyon Lavochkin)
HEND	High-Energy Neutron Detector

ICBM	InterContinental Ballistic Missile
IKI	*Institut Kosmicheski Izledovatl* (Institute for Space Research)
IMAP	Magnetometer
IMBP	Institute for Medical and Biological Problems
IRE	Institute of Radiotechnology and Electronics
IZMIRAN	Institute of Terrestrial Magnetism
JIMO	Jupiter Icy Moon Orbiter
KAMERTON	Seismometer
KBKhA	*K.B. Khim Automatiki* (Design Bureau for Chemical Automatics)
KGB	Internal security police
KMV	*Korabl Mars Venera* (spaceship for Mars and Venus)
LAL	Long-term Automated Lander
M	Modified version (e.g., 8K78M)
MAREMF	Electron spectrometer and magnetometer
MARIPROB	Plasma and ion detector
MARSPOST	MARS Piloted Orbital STation
Mavr	Linguistic combination of Mars and Venera
MEK	Mars Expeditionary Complex
METGG	Meteorology instrument system
MPK	Mars Piloted Complex
MV	Mars Venus
N	*Nositel* (carrier)
NASA	National Aeronautics and Space Administration
NEK	Scientific Experimental Complex
NII PDS	Scientific Research Institute of Parachute Landing Facilities
NIPs	Scientific measurement points
NITS	Babakin Scientific Research Centre
NPO AP	Scientific Production Association for Automatics and Instrument Development
NPO	Science and Production Organization
NPO-PM	Scientific Production Association *Pridlanoi Mekhaniki*
NSSDC	National Space Science Data Centre
OIMS	All-Union Society to Study Interplanetary Communications
OKB	*Opytnoye Konstruktorskoyue Buro* (experimental design bureau)
OPTIMISM	Seismometer, magnetometer
OSOVIAKHIM	Society for the Support of Aviation and Chemical Development
P	*Pasadka* (lander)
PEGAS	Gamma-ray spectrometer
perestroika	Transformation, reform
PHOTON	Gamma spectrometer
PrOP-M	*Pribori Otchenki Prokhodimosti-Mars* (instrument for evaluating cross-country movement)

PS	Preliminary Satellite
RKA	Russian Space Agency
RT	*ReTranslyator* (relay station)
SFINCSS	AKA *Sphinx*
SLED	Solar Low-Energy Detector
SPICAM	Optical multichannel spectrometer
SVET	High-resolution mapping spectrophotometer
TERMOSCAN	Scanning infrared radiometer
TERMOZOND	Temperature probe
TMK	*Tizhuly Mezhplanetny Korabl* (Heavy Orbital Station)
TsAGI	*Tsentralni Aero Girodinamichevsky Institut* (Zhukovsky Central Institute of Aerodynamics)
TsBIRP	Central Bureau for the Study of the Problems of Rockets
TsDUC	Centre for Long Range Space Communications
UDMH	Unsymmetrical Dimethyl Methyl Hydrazine
UMV	Universal Mars Venus
UMVL	Universal Mars Venus Luna
UR	*Universalnaya Raketa* (Universal Rocket)
VEGA (VEHA)	VE = Venus and HA = Halley ('g' and 'h' sound similar in Russian)
VPM-73	Visual Polarimeter Mars 1973

1

Aelita

That Mars is inhabited by beings of some sort or other we may consider as certain as it is uncertain what those beings may be.

– Percival Lowell, in *Mars and its canals*, 1906.

THE LONG SCHOOL SUMMER OF 1883

Russia's exploration of the planets dates to summer of 1883 and the writings of a teacher in the town of Kaluga, Konstantin Tsiolkovsky. He was the first Russian to describe interplanetary journeys.

Born in Izhevskoye in September 1857, Konstantin Tsiolkovsky was the son of a forest ranger, but suffered from scarlet fever when he was ten years old and this left him largely deaf for the rest of his life. His mother nursed him through school and, possibly on account of his deafness, he read books endlessly, by the age of 17 mastering higher mathematics, differential calculus and spherical trigonometry. He managed to obtain a teaching post in the town of Kaluga, where he spent the rest of his life.

During the long school summer of 1883, Tsiolkovsky came across the idea of interplanetary space travel, though we still do not know precisely how or exactly what sparked his interest. This became the guiding interest of his life. It should be noted that Tsiolkovsky was no dreamy theorist. This remarkable self-taught man was also a mathematician, inventor, writer and and practical engineer, for he built hearing aids, wind tunnels and even small centrifuges.

That summer, 1883, Tsiolkovsky wrote his first book, *Free space*, which described how an interplanetary traveller would be weightless. He went on to devise the formula for the thrust of a rocket engine and 'Tsiolkovsky's formula' became the basis of all

Konstantin Tsiolkovsky

subsequent rocket science and is now the first thing taught in rocket school. In *On the moon* (1893) and *Dreams of Earth and heaven* (1895) he outlined flight beyond the Earth – accurately calculating the velocities required – and in *Interplanetary flight* he sketched a space station that could travel deep into space bringing with it its own, closed ecosystem. *Investigating space with reaction devices* (1903) explained how decaying nuclear fuel would be necessary for long space journeys, for they offered superior propulsion to liquid-fuel rockets. He showed how it might be possible to soft-land on other planets. Eventually, he predicted, 'people will ascend into the heavens and found settlements there.'

In *Cosmic rocket trains* (1918) Tsiolkovsky turned specifically to the problem of reaching a sufficient velocity to escape Earth's gravity for interplanetary journey. The key was to use rocket stages, with one stage falling away after another, either on top of one another or in parallel, each achieving ever greater height and speed than its predecessor. In a famous letter to B.N. Vorobyov on 12th August 1911, Tsiolkovsky wrote:

Human kind will not remain forever confined to Earth. In pursuit of light and space it will, timidly at first, probe the limits of the atmosphere and later extend its control across the entire solar system.

The Bolshevik government gave him a pension and in the final years before his death in September 1935, his home in Kaluga became a place of pilgrimage for amateur rocket-builders, enthusiasts, journalists and popularizers of science. He published no fewer than 60 works – articles, books, even science fiction – in his last ten years. He lived out his last autumn days on his veranda, surrounded by books, manuscripts and the odd globe. His home is now a shrine, museum and memorial.

AELITA AND THE ROLE OF SCIENCE FICTION IN RUSSIA

Science fiction circulated widely in pre-revolutionary and revolutionary Russia, the writings of Frenchman Jules Verne being much the most popular. One of those

Jules Verne – popular in tsarist Russia

inspired by Jules Verne was a young boy from Odessa on the Black Sea, Valentin Glushko. When aged 13 in the year 1921, he read the 1865 book by Jules Verne *From the Earth to the moon*. In the following year, still only 14, he made his own observations of Venus, Mars and Jupiter in the Odessa Observatory and these were published the following year in *Astronomical Bulletin* and the magazine *Mirovodenie*. The following year, aged 16, he wrote to and received a treasured reply from Konstantin Tsiolkovsky. When Valentin Glushko completed his studies in 1928 he went straight into the Gas Dynamic Laboratory (GDL) in Leningrad to design and build rocket motors, becoming, eventually, the greatest rocket engine designer of all time [1].

Verne's science fiction may have been the first of its kind in Russia and it unleashed a huge upsurge in the genre. Mars was the focus of Russian science fiction. In the same year that H.G. Wells published *War of the worlds*, 1908, the dissident Bolshevik, philosopher and doctor Alexander Boganov (1873–1928) published *Red Star* (*Krasnaya Zvezda*), the tale of benign Martians seeking cooperation from humans. The landmark novel of the period was *Aelita*, written in 1923 by Alexei Tolstoy (1883–1945). It was the story of two cosmonauts who arrived on Mars to find it under the yoke of a cruel empire. To complicate things, one of the two, cosmonaut Los, fell in love with Princess Aelita, Queen of Mars, after whom the novel was named. Meanwhile, fellow cosmonaut Gusev led a plot to overthrow her imperial government and establish a model society. The novel was approved by the authorities and quickly turned into a popular film, in turn prompting interest in the written text. The film, directed by the recently returned from exile Yakov Protazanov, had a cast which included Tuskub, King of Mars and a 'bearded astronomer' and was rated highly for its design and production values. Yuliya Solntseva (1901–89) starred as Queen Aelita, setting her off on a long career in acting and, after the war, as a director, winning international recognition (she was on the jury of the Cannes film festival). *Aelita* had such an appeal that some parents named their daughters Aelita.

The success of *Aelita* later inspired a second film, the silent movie *Space journey* (1935), a flight to the moon which showed cosmonauts weightless and in spacesuits, the first ever film to do so (the ailing Tsiolkovsky was technical advisor). A novel of the time also dealt with a flight to Mars: *Jump into nowhere*, a less political novel describing a scientific mission to Mars on a liquid propellant rocket.

The 1920s were a period of space fever in the new Soviet Union [2]. The new government was quick to enlist technology in support of national industrial, economic and scientific development. Aeroplane design was used as a cutting edge design field to improve Soviet technology and a vast aeroplane, the *Maxim Gorky*, was constructed as an airborne propaganda unit. Manned balloons flew high into the atmosphere. As early as 1918, Lenin authorized the establishment of the central institute of aerodynamics, TsAGI (*Tsentralni Aero Girodinamichevsky Institut*) by the founder of Russian aviation, Professor Nikolai Zhukovsky. The following year saw the establishment of the Zhukovsky Academy. The year 1924 saw the setting up of the Central Bureau for the Study of the Problems of Rockets (TsBIRP) which saw its brief as 'to disseminate and publish correct information about the position of interplanetary travel' and also the All Union Society to Study Interplanetary Communications (OIMS), whose title was self-explanatory, with 200 members who included Tsiolkovsky and Tsander. Its first meeting was in Pulkhovo Observatory on 20th June 1924. Interplanetary clubs appeared in other cities. A Society for the Support of Aviation and Chemical Development, OSOVIAKHIM, was formed to enlist and develop public support for aviation, including rocketry and it attracted wide public support.

PERELMAN, TSANDER AND SHARGEI

The greatest popularizer of spaceflight in the early years was Yakov Perelman (1882–1942) who published – in 1915 wartime Petrograd – *Interplanetary travel*. This was so popular that it went through ten reprints in the following 20 years. *Interplanetary travel* explored how the idea of reaching distant planets could be made possible through such means as anti-gravity (H.G. Wells), a great gun (Jules Verne), solar propulsion or Tsiolkovsky's rocket, which he favoured. Perelman's *Interplanetary travel* was republished in 1929 in an entirely new edition and at once sold 150,000 copies. Perelman became vice-chairperson of LenGIRD, the Leningrad branch of the

Yakov Perelman

amateur Group for the Study of Reactive Devices (GIRD), whose Moscow branch was to launch the first liquid-fuel rocket. Between 1917 and 1941, no fewer than 535 separate publications addressed issues of spaceflight.

Riga-born Lithuanian Friedrich Tsander (1887–1933), a son of a merchant family on his father's side and musicians on his mother's part, was read Tsiolkovsky's *Exploration of space by reactive devices* by a teacher in the course of school in 1905 and thereafter resolved to devote his life to space travel. His house stood on a hill, and there he made a shed from which he could observe the moon, Saturn and Mars. He made mathematical calculations as to the best paths of interplanetary journeys, landings and return trajectories [3].

Graduating from Riga Polytechnic, Friedrich Tsander went to work in the *Motor* aircraft plant in Moscow in 1919. In his spare time, he gave lectures on space travel in the early Soviet Union and encouraged many people to join the cause. Even when Moscow was freezing and starving, he set up a model hydroponic garden to grow food in an air-tight laboratory, simulating the conditions of interplanetary travel. Indeed, his children were enlisted in the good cause, for he named his daughter Astra and his son Mercury. To Tsander may be attributed the first ideas of a space shuttle, for he saw that winged spaceships were an efficient way to land on other worlds that had an atmosphere and a means of returning to the Earth. He outlined his concept of a space shuttle and the use of aircraft to bring a rocket to altitude in lectures to the Moscow regional conference of inventors in 1921. His main work was *Flights to other planets*, published in the magazine *Technology and life* in 1924. He was no mere theorist, for he joined the GIRD group and spent the last years of his life building small liquid-fuel rockets. In the course of doing so and aged only 46, he contracted typhoid fever. His slogan in GIRD was 'Onward to Mars' because it was the planet with an atmosphere and the place most likely to sustain life. In the view of the GIRD leader, Sergei Korolev, it was Friedrich Tsander more than anyone else in the group who looked furthest ahead to interplanetary flight. Tsander died in Kislodovsk Sanitorium in the

Friedrich Tsander: 'Onward to Mars!'

Yuri Kondratyuk

Caucasus on 28th March 1933. Korolev broke the news to the entire, shocked GIRD group in its basement laboratory. Korolev concluded his speech at the funeral with Tsander's own words, 'Onward to Mars!'

It was Yuri Kondratyuk (1897–1942) who wrote the most about the potential of interplanetary travel during the early days. Yuri Kondratyuk was mentioned fleetingly in the Soviet histories of the 1950s and years later the enigma surrounding this extraordinary man began to unravel. Kondratyuk was not his name at all, for he was really Alexander Shargei. During the civil war, he was conscripted into the White army, the Bolshevik penalty for which was death. To survive, he deliberately took on the identity of a dead man, Yuri Kondratyuk, acquiring his birth certificate.

In 1916, while still under his original name of Shargei, he wrote *Conquest of interplanetary space* in the polytechnic in Petrograd, but circulation was restricted during the wartime conditions and he was soon forced into the tsar's army in any case. This was revised in 1919 as *To those who will read to build*, but its circulation was overtaken by the civil war now raging. *Conquest of interplanetary space* was eventually published in 1929, but as a new, first edition, now under the name of Kondratyuk, in the hope that memory of the now receding 1916 edition would be forgotten. In an additional precaution to avoid the attention of the authorities, he had it published in distant Novosibirsk. *Conquest of interplanetary space* was important, for it outlined how a rocket could leave Earth and fly into space, the use of decaying atomic fuel and solar energy for propulsion and addressed issues of trajectories, guidance and stability. For landing on another planet, he recommended that the bottom stage be left behind and serve as a platform for the return spacecraft, the very technique followed by the American and Russian lunar modules 40 years later. He also suggested that, instead of direct landings, spacecraft be sent down from orbit around the planet, to which they would return for their flight home. For returning to the Earth, he recommended either aerobraking or a high-speed entry into the atmosphere, the very approach followed by Zond and Apollo to return from the moon over 1968–72. Shargei suggested the use of gravity assist to bounce from one planet to another, a technique later used by the Russian VEGA spacecraft.

Kondratyuk was several times invited to join Friedrich Tsander's rocket group, but in letters he kept pushing them away, privately afraid that if he were to be scrutinized by the police for a new job with them, his secret identity would be exposed. He eventually and probably reluctantly left Moscow to get away from them, become a hydro-electric plant engineer and died fighting the Germans in the snows outside Moscow in February 1942. German troops picked up Yuri Kondratyuk's notebook, full of interplanetary drawings, from his body. But his true identity was not uncovered for more than half a century.

When John F. Kennedy decided the United States should go to the moon in 1961, NASA spent almost 18 months deciding on the best way to get there. Both direct ascent using a huge rocket and Earth orbit rendezvous presented significant problems. The deadlock was only broken where someone remembered the theories of Yuri Kondratyuk for rendezvous in planetary (or lunar) orbit, the technique eventually adopted.

So high was the level of interest in space travel in the 1920s that Professor Nikolai Rynin made a nine-volume compilation of all that was known about the subject at the time. Called *Space travels*, it was in reality an encyclopaedia. Rynin had been organizer of Russia's aerodynamic laboratory in the days of the tsar. Not much aerodynamics was undertaken during the revolution or subsequent civil war, so Nikolai Rynin began his compilation quietly during this difficult time, even though he was cold and hungry. Only one Soviet space encyclopaedia was ever published subsequently, by Valentin Glushko in 1985.

The first exhibition of spacecraft designs was held in Moscow from April to June 1927 under the auspices of the Association of Inventors and it featured the works of Tsiolkosvky and Tsander as well as famous westerners. 12,000 people came to see the models and drawings. Exhibitions were also held in Kiev.

AFTER SUPPRESSION, REVIVAL

The Soviet space boom ended in 1936–7. Stalin decided that the construction of socialism required a strict focus on national industry and military defence, a world in which space travel, collaboration with foreign organizations and amateur societies had no place. The patron of the main rocket research institute, Marshal Tukha-chevsky, was shot, as were two leading rocketeers Ivan Kleimenov and Georgi Langemaak. GIRD leader Sergei Korolev was sent to the Gulag and then worked in a confined location (*sharashka*), while the GDL's Valentin Glushko was soon arrested too. The chairman of OSOVIAKHIM, Robert Eideman, was executed.

Rocketry and space development in the USSR temporarily ceased in 1937, although some limited work was undertaken on the development of small solid-fuel rockets to help to get military aircraft airborne quickly from short runways. Both Yakov Perelman and Nikolai Rynin died of starvation during the siege of Leningrad.

Not until Stalin's death was the discourse evident in the first 15 years of the Soviet Union restored. Fiction and non-fiction writing about space exploration came out into the open once more. A complete collection of Tsiolkovsky's works was published in 1954, as were the works of Tsander and Kondratyuk. Some of this was inspired by the desire, at a time of American military supremacy, to show that Soviet science was

МЕЖПЛАНЕТНЫЕ СООБЩЕНИЯ

Н. А. РЫНИН

ЛУЧИСТАЯ ЭНЕРГИЯ

1 · 9 · 3 · 1

Nikolai Rynin's book

as good as that of the West. Mars fiction reappeared with Georgi Martinov's *220 days in a starship*, in which American astronauts travel first to Mars, but they become stranded and must be rescued by Russian cosmonauts. A conference took place in Leningrad in February 1956 to discuss the Moon, Mars and Venus. The Soviet media of the 1950s were full of discussion of future flights of satellites and, later, journeys to the moon and nearby planets. By 4th October, Soviet citizens were well prepared for the first satellite launching. They were not in the slightest surprised, unlike ordinary American people, though this did not dim their pride in its achievement. Sergei Khrushchev, son of the Soviet leader, pointed out that his generation was reared on the science fiction of Tolstoy and Perelman. The launch of an Earth satellite was not a fantastic idea at all: 'we had been waiting for engineers to turn the dream into reality' for some time [4].

VENUS AT THE TIME OF THE SPACE AGE

And what of the planets that they hoped to visit? Venus was well known to the ancient astronomers. The Italian Galileo Galilei was the first to observe the phases of Venus

Crescent Venus through a telescope

through his telescope in 1610. It was a Russian, Mikhail Lomonosov (1711–1765), who was the first to determine in 1761 that there was an atmosphere around the planet and estimated that it to be at least equal to and possibly greater than Earth's atmosphere [5].

The early telescope astronomers found Venus a really frustrating object. Most quickly came to the conclusion that the planet was shrouded in thick cloud, though this did not stop some astronomers with gifted eyesight and observational powers from making and publishing maps of the planets showing its landmasses, continents and oceans. Venus was calculated to be a similar size to Earth, which it is, 12,104 km in diameter compared with Earth's 12,756 km. Venus orbits at an average of 108.2m km from the sun, compared with Earth's 149m km. The clouds made the rotational period difficult to calculate and the matter was not settled until the late 20th century (243 days). An important breakthrough came in the 1930s, when observers at the Mount Wilson Observatory in the United States fitted filters to their telescopes in an attempt to characterize the atmosphere of the planet. They found a high level of carbon dioxide, with suggestions of a high surface temperature. Up till this time, the popular image was that – just as Mars was a dry world, with the Martians conserving its remaining water in its canals – whilst in symmetry, Venus was a dripping wet, steaming, swampy, carboniferous planet, like the jungle of the Congo, such a picture being painted in 1918 by Swedish Nobel Chemistry prize winner Svane Arrhenius.

Enter Gavril Adrianovich Tikhov (1875–1960). Although hardly known in the West, he had a profound influence on how Venus (and Mars) were perceived at the time of the space age and, in turn, on the direction of the early exploration programme there. He was born in Smolevichi, Minsk on 1st May 1875, the son of a railway station master. He was a promising child, winning a gold medal when he graduated from Simferopol Gymnasium in 1893, whence he went to Moscow University to study

Gavril Tikhov

mathematics for the following four years. Graduating from there in 1897, he married Ludmila Popova and then travelled to France – and from 1898 to 1900 enlisted in the Sorbonne. There, astronomy captured his attention and he became assistant to Jules Janssen at Meudon Astrophysics Observatory. It was there that Gavril Tikhov first showed that astronomy was not something to be done only from books and began to show his practical bent. He ascended in a balloon with fellow Russian astronomers A.P. Gansky and French colleagues on 15th November 1899 to observe meteors and later went up Mont Blanc to get a better view of the heavens.

Gavril Tikhov returned to Moscow where he earned an income as a mathematics teacher. Not long after, he was invited to Pulkhovo Observatory in St Petersburg by the astrospectroscopic pioneer A.A. Belopolsky and began to study for an MA there in astronomy and geodesy, which he was awarded in 1913. It was in Pulkhovo that he first began to study the planets in 1909 with a 76.2 cm telescope. Belopolsky pioneered the use of coloured filters to improve our knowledge of stars, but Tikhov applied them to the surface of Mars.

The war put an end to his astronomical studies and in 1914 he joined the Central Aeronautical Navigation Station. There, he applied his observational skills to develop aerial photography to be used for military photoreconnaissance. The systems he developed were well regarded and used by the British and French Air Forces. He wrote the first textbook on aerial reconnaissance, *Improvement of photography and visual air intelligence*, before being conscripted into the army in 1917.

The war over, in 1919 Gavril Tikhov returned to work at Pulkhovo Observatory, work which he combined with a lectureship in the University of St Petersburg, later Leningrad. He was appointed Director of Astrophysics, a post he held till 1941. In the meantime, he was elected to the Academy of Sciences in 1927 and was awarded a PhD in 1935. In 1941, he headed for central Asia to observe the eclipse there and ended up staying in the region. Many leading scientists were evacuated there during the war.

Once the war was over, the government asked him to organize the Kazakh Academy of Sciences. This suited him, for although most of evacuated scientists were happy to return to Moscow and other regions of Russia, some were content to remain. Alma Ata became one of the leading centres of excellence in Soviet astronomy.

Gavril Tikhov in Pulkhovo

There in 1947 he established the world's first department of astrobotany, which he headed (it would now be called exobiology). His observatory was equipped with a 20 cm Maksutov telescope specially adapted for lunar and planetary observations (a 70 cm replacement arrived in 1964). Central to Tikhov's approach was the notion that plants were plentiful on the planets and he believed that he could predict the nature of such plant life through observation. Whilst some of his claims may now be regarded as fanciful, Tikhov was a serious observer and his skills had already been successfully applied to aerial photography. His observatory attracted young graduates from all over the Soviet Union. He often led them on expeditions into the high mountains to study how plants survived in cold, high-altitude, thin-air régimes. In his lectures, Tikhov was a declared enemy of geocentrism (that life in space should be imitative of Earthly life) and argued that it could take many different forms. According to a writer on Tikhov, David Darling, the idea of life on Jupiter's icy moons would not have surprised him [6].

Gavril Tikhov on astrobiology

Tikhov was an acknowledged expert in the colours of stars and variable stars. He was a voluminous writer and popularizer of astronomy, publishing no fewer than 165 books, articles and papers, including an autobiography *Sixty years at the telescope* (1959). His most famous books were *Astrobotany* (1949) and *Astrobiology* (1953). Turning his attention to Venus, Tikhov analyzed, through filters, the light of Venus. Due to the nature of its atmosphere, the plants there, he predicted, would be blue plants rather than Earth's homely green.

As the space age drew near, Tikhov inspired a renewed interest in the planets in the USSR. Telescopic studies of Venus intensified in the 1950s, notably by Russian astronomer N.A. Kozyrev. The views of Venus were still benign. Surface temperatures were estimated in the region of 60°C to 76°C with watery oceans.

Tikhov's conventional wisdom was challenged by V. Beloussov of the Academy of Sciences who predicted in 1961 that Venus would be found to be a hot, primaeval, lifeless planet full of ridges and volcanoes. The then vice-president of the Academy of Sciences Mstislav Keldysh asked the radio astronomers of Pushino Observatory to use their 22 m dish to scan Venus' surface to try determine its temperature. Similar experiments were carried out in the United States and in other parts of the world, including collaborative American – Russian ones, but the results were unsatisfactory, variable and inconclusive, suggesting surface temperatures in the range 300°K to 600°K, which was quite a big difference (K is the measurement of absolute zero, or −274° Celsius).

The heretical scientist I.S. Shklovsky, who once infamously declared that Mars' moons Phobos and Deimos were artificial, predicted that Venus would be murky, red hot, no water, a dense atmosphere of carbon dioxide with high surface temperatures (300°C) and pressures [7]. When Sergei Korolev first designed his Venus probes in the late 1950s, he assumed a pressure of 1.5 to 5 atmospheres, temperature of up to 75°C and a composition of carbon dioxide and nitrogen. As late as 1966, Soviet astrophysicist Nikolai Barabashov was still holding to the view that there would be oceans on Venus, 'the cradle of life, possibly at a level of development which existed on Earth millions of years ago' [8].

MARS AT THE TIME OF THE SPACE AGE

Mars was known to the ancient Egyptians and Chinese and was documented by Aristotle and Ptolemy. Its movements were followed by the great astronomers Brahe and Kepler, with the first drawings of the planets undertaken by Christian Huygens in 1659. By the 17th century, the main characteristics of its orbit had become known and some of the observations of the early astronomers were astonishingly accurate. Mars had an equatorial diameter of 6,794 km, a rotation period of 24 hr 37 min, took 687 Earth days to orbit the Sun, which it circled between 207m and 249m km (average: 228m km).

Lacking the type of clouds that obscured Venus, Mars at least offered possibilities for observers, so many astronomers tried to map the planet. Earth and Mars would pass relatively close to one another every two years ('opposition') and some of these oppositions would be quite close, providing favourable opportunities for observation. During such approaches, the planet's reddish colour was quite evident and it was possible to make out the planet's polar caps and darker and lighter areas on the surface. As early as 1770, Earthly astronomers noted its 'fiery red colour' and that it seemed 'to be encompassed by a very gross atmosphere' [9].

The first recognizable modern maps were published by Giovanni Schiaparelli in 1877 and the period of pre-space age Martian mapping was brought to a conclusion by Henri Carmichael at the Pic du Midi Observatory in France, with the maps compiled being approved by the International Astronomical Union in 1958 [10]. So, unlike Venus, Earthlings had some idea of what Mars might be like and these ideas could be confirmed (or refuted) by space probes.

The person most indelibly associated with the astronomy of Mars was American astronomer Percival Lowell. He claimed to have found an elaborate system of canals, which he quickly came to the conclusion must have been built by an advanced civilization (which, he speculated, was in its final stage, desperately trying to conserve the water of the receding Martian poles). He publicized his theories in *Mars* (1895),

Mars and its canals (1906) and *Mars as the abode of life* (1908). It is no coincidence that H.G. Wells' *War of the worlds* was published in the same year as the last, sparking off a round of Martian science fiction revolving around good Martians and bad Martians that has yet to run its course. Although most scientists were slow to accept the idea of a red planet inhabited by a civilization, the idea of an inhabited planet with the possibility of some form of life became firmly embedded in the popular mind and that became Lowell's historic legacy. Not long before the first space probes arrived there, the National Academy of Sciences in the United States adjudged that it was 'entirely reasonable that Mars was inhabited with living organisms'.

Gavril Tikhov had been observing Mars from Pulkhovo Observatory in Leningrad since the 1930s. The colours of Mars – red, brown, green – encouraged him to compare them with how the colours of Earth would look from space. Mars, he figured, displayed the same colours as the Soviet Arctic, with its snows, mountains, soils and vegetation. Ergo, Mars might be similar to Earth's northern latitudes. He undertook a spectral analysis of seasonal changes on Mars and, in the mid-1950s, toward the end of his long life, he was invited to give public lectures on the theme *Is there life on Mars?* Tikhov may be ridiculed for his Venus plants now, but his general approach was less wide of the mark than he was given credit for. He took the view, unfashionable at the time, that life could survive in conditions that were not necessarily Earthlike. Even on Earth, there were plants that survived without oxygen, preferring ammonia. He pointed out that a Martian would regard life as impossible on Earth, because we would all smother in oxygen. He argued that life could be possible on Mars, just as life could be found on cold, high plateaux on Earth. Tikhov travelled thousands of kilometres across southern Russia and to the Siberian Arctic to show the extreme conditions in which life could survive and thrive, taking 15 expeditions there. Accordingly, finding life on the planet became a dominant, early theme, in Russian Mars exploration. Tikhov lived to be 85 and died in Alma Ata, Kazakhstan on 25th January 1960.

In the event, his department was not to shape the practical design of instruments for the first probes to Mars and Venus. Here, the Vernadsky Institute became ascendant [11]. Located in central Moscow, the institute comprised many of the scientists who designed the instruments for the first lunar, Venus and Mars probes. From 1961, the Vernadsky Institute had a planetary geochemistry laboratory, headed by one of Russia's most famous scientists, Yuri Surkov, who shaped the instrumentation programme and wrote extensively about the results.

The conventional impression of Mars in the early to mid-1960s was that it was a relatively benign world. Atmospheric pressure was estimated to be in the order of 80 to 120 mb and this included the findings of Soviet astronomers such as Barabaschev and Sytinskaya (116 mb was the favourite quoted figure). The *Encyclopaedia of Astronautics* gave it an atmospheric pressure of 65 mm and balmy, Californian temperatures [12]. Although artificial waterways were no longer seriously considered, there was broad endorsement of the view that the spring melting of polar caps prompted seasonal vegetation to blossom. The atmosphere was so turbulent and erosive that there could be no possibility of high mountains. When Korolev first designed his Mars probes, he worked on the assumption that the temperature of Mars was in the order of

Yuri Surkov

$-70°C$ to $+20°C$ and that the thin but useful atmosphere comprised mainly nitrogen and some oxygen.

In his chapter *Mars before the space age* in his book *On Mars*, astronomer Patrick Moore relates how he used to give lectures on Mars in the University of London in the early 1960s and that almost everything he said subsequently turned out to be wrong. Theories of the canals of Mars as evidence of advanced civilization had never won wholehearted acceptance in the scientific community, but a benign view of the planet was endorsed. The atmosphere appeared to be rich in water vapour. The main component was estimated, like Earth's, to be nitrogen. All these factors pointed to a world that could support at least low-level life [13].

PRELUDE TO THE INTERPLANETARY AGE

The Soviet Union was therefore well prepared for the first flights to the planets. Thanks to foreign writers such as Jules Verne and domestic ones such as Yakov Perelman, there was a long history of science fiction in Russia, one in which travel to the planets was a recurrent motif and brought to a mass audience by films such as *Aelita*. The space boom in the 1920s and early 1930s meant that the idea of flight to the planets was quite broadly understood in Soviet society. The theoretical basis for flights to the planets had been set down with impressive foresight by the writers Tsiolkovsky, Tsander and Kondratyuk, people who combined imagination with a practical grasp of physics, motion and propulsion. Russian astronomers had, like the rest of the world-wide astronomical community, interested themselves in Venus and Mars, finding and measuring Venus' atmosphere as far back as the 18th century. Gavril Tikhov had developed the idea of astrobotany, putting the finding of surface or subsurface life top of the exploration agenda.

REFERENCES

[1] For an early history of this period, see Riabchikov, Yevgeni: *Russians in space*. Weidenfeld & Nicolson, London, 1972.

[2] Gorin, Peter A.: Rising from a cradle – Soviet perceptions of spaceflight before Gagarin, in Roger Launius, John Logsdon and Robert Smith (eds): *Reconsidering Sputnik – forty years since the Soviet satellite*. Harwood, Amsterdam, 2000.

[3] One day we shall fly to Mars. *Soviet Weekly*, 27th August 1977.

[4] Khrushchev, Sergei: The first Earth satellite – a retrospective view from the future, in Roger Launius, John Logsdon and Robert Smith (eds): *Reconsidering Sputnik – forty years since the Soviet satellite*. Harwood, Amsterdam, 2000.

[5] Moore, Patrick: *The Guinness book of astronomy*, 5th edition. Guinness Publishing, Enfield, UK. 1995.

[6] Darling, David: *Gavril Tikhov* at *http://www.daviddarling.info.encyclopedia/t/tikov*

[7] Burchitt, Wilfred and Purdy, Anthony: *Gagarin*. Panther, London, 1961.

[8] Russians contact planet Venus. *Irish Times*, 2nd March 1966.

[9] Ferguson, James: *Astronomy*. Strahan & Co., London, 1770.

[10] Corneille, Philip: Mapping the planet Mars. *Spaceflight*, vol. 47, July 2005.

[11] Hansson, Anders: *V.I. Vernadsky, 1863–1945*. Paper presented to the British Interplanetary Society, 2nd June 1990.

[12] De Galiana, Thomas: Concise Collins *Encyclopaedia of Astronautics*. Collins, Glasgow, 1968.

[13] Moore, Patrick: *On Mars*. Cassell, London, 1998.

2

First plans

One day we shall fly to Mars.

  – Friedrich Tsander, 1932

SOVIET ROCKETRY

The theoretical basis for interplanetary flight had been well laid by Russia in the period from 1883, Tsiolkovsky's first book, to the space boom of the 1920s and a resumption of interest in space travel in the 1950s. What about the practical side?

The 1920s and 1930s saw the development of rocketry in the USSR by both professionals and amateurs. Professional rocket development was undertaken in the Gas Dynamics Laboratory (GDL) in Leningrad from 1928. There, over the next ten years, thousands of firings were made of small liquid-fuel rockets. GDL was the place of the apprenticeship of Valentin Glushko, who built and static-tested his first small engine, the ORM-1, in 1931. Glushko went on to experiment with different types of fuels: not just the conventional ones, like kerosene, but nitric acid and other fuels that ignited on contact with one another and he pioneered turbine pumps, cooling systems and throttles. By 1933, 200 staff were working in GDL.

The first rockets to be actually fired were constructed by amateurs in Moscow called GIRD, the Group for the Study of Reactive Propulsion. The leader of this group was the man who became the greatest of all Soviet space designers, Sergei Korolev. A contemporary of Glushko, Sergei Korolev was born in Zhitomir in the Ukraine two years earlier in 1906. Korolev was a graduate of Kiev Polytechnic and Moscow Higher Technical School, becoming a designer, developer and pilot of rocket-propelled gliders. But his real interest was rocketry and he assembled teams of like-minded companions into GIRD; they would go into Moscow's forests over the

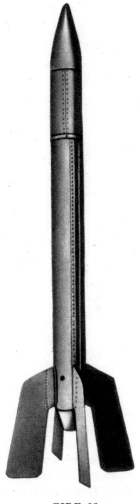

GIRD-09

weekends to test their latest experiments. His close GIRD colleague, Mikhail Tikhon-ravov, later a driving force in the early moon programme, built the 2.4 m long, 19 kg GIRD-09 which on 17th August 1933 took off above the surrounding conifer trees and was the first small rocket ever fired in Russia. But with the purges three years later, all this came to an end.

GERMANY'S PROGRESS

The first modern rocket was launched not by Russia, but by Germany. The momen-tous happening took place on 3rd October 1942, when a German A-4 rocket left its base at Peenemünde and flew in a high arc 190 km over the Baltic, splashing down in the sea 15 min later. It was larger, more powerful and sophisticated than anything that had flown before: by way of comparison with the little GIRD-09, the A-4 was 14 m high and weighed 12.8 tonnes with a payload of a tonne. The A-4, designed by Wernher von Braun, was subsequently used to bombard London, Antwerp and other cities with devastating results.

Russian engineers were taken aback at the pace of Germany's rocket advances. In August 1944, on a tip-off from British intelligence, Mikhail Tikhonravov was dispatched to Debica, Poland where the remains of a crashed German A-4 had been identified by Polish partisans. They found the engine and other parts, dredging them from out of a swamp whence they were shipped back to Moscow. The Russians were amazed at the quality of the workmanship, the size of the engine and how the Germans had managed to overcome problems of guidance and control.

When the Red Army captured the Baltic coast in 1945, Soviet engineers followed quickly behind to pour over the remains of the A-4 and Germany's other wartime rocket achievements. An undignified scramble took place to loot the scientific heritage of the Reich, with the Russians competing with the Americans, British and French and all of them with each other. Sergei Korolev was speedily released from his *sharashka*[1] in October 1945, followed by Valentin Glushko. Both even attended the launch of a captured A-4 by the British from Cuxhaven, Germany. The Peenemünde Germans captured by the Red Army were put to work by the Russians and then transferred, with their families, to a guarded island in Seliger Lake between Moscow and Leningrad. At Peenemünde the Russians found, to their astonishment, the notebook of Yuri Kondratyuk, picked from his body in the Moscow snows in 1942.

Key developments soon followed, all driven by Stalin's desire to have a long-range military rocket. A Council of Chief Designers was set up in 1946, led by newly appointed chief designer Sergei Korolev. A design bureau was formed to lead the programme, called OKB-1 (*Opytnoye Konstruktorskoye Buro*) (Experimental Design Bureau). A cosmodrome was quickly constructed east of Volgograd, called Kapustin Yar. German A-4s were fired from there from October 1947 and a reverse-engineered Russian version, called the R-1, in October 1948.

R-7 ROCKET

In December 1950, Korolev was asked by Stalin to build an intercontinental ballistic missile (ICBM), able to deliver a nuclear weapon across continents to the United States. It was the seventh he designed and acquired the title R-7 (R for *Raket*, or rocket in Russian). It is worth giving further detail at this stage, for the R-7 was to play a pivotal role in the Mars programme until end 1964 and in the Venus programme until 1972. An ICBM in the 1950s was a step beyond the A-4, as much as the A-4 of the 1940s was a step beyond the tiny amateur rockets of the 1930s, like the GIRD-09. Korolev was the mastermind of what became known as the R-7 rocket. It was larger

---

[1] A sharashka is a form of house arrest.

Sergei Korolev

than any rocket built before. It used a fuel mixture of liquid oxygen and kerosene, a significant improvement on the alcohol fuel used on the German A-4. Powerful engines were designed and built by the GDL's Valentin Glushko, who now had his own design bureau, OKB-486.

The real breakthrough for the R-7 was that – in addition to the core stage with four engines (block A) – four stages of similar dimensions were grouped around its side (blocks B, V, G and D). This was called a 'packet' design. Originally an idea of Tsiolkovsky, it was refined by Mikhail Tikhonravov in 1947. No fewer than 20 engines fired at liftoff. The side stages (B, V, G, D) dropped off two minutes into flight, block A bringing the payload into orbit. Korolev and Tikhonravov designed the R-7 to have a thrust of 600,000 kg, the ability to deliver a one-tonne military payload on an enemy or an orbital payload of 1,350 kg. The R-7 was delivered horizontally to its pad on a huge railcar. It was then tilted upward and set in the restraining arms of the pad. A minute from launch, the arms swung back. After ignition, once the rocket reached sufficient thrust, it simply lifted off and the remaining clamps, their burden released, swung back. It was important to hide the construction of an intercontinental ballistic missile from the Americans, so a new launch site was built in the deserts of Kazakh-stan. It was called Baikonour, though the real Baikonour was a sleepy railhead 370 km distant: presumably in a nuclear war, the Americans would bombard the hapless folk of the real Baikonour, while the 'Baikonour' cosmodrome survived.

R-7 rocket

Length	33.5 m
Diameter	10.3 m
Weight	279.1 tonnes
of which frame	26.9 tonnes
propellant	256.2 tonnes
Thrust at liftoff	407.5 tonnes
Burn time block A	320 sec
Burn times blocks B, V, G, D	120 sec

THE IDEA OF AN EARTH SATELLITE

The early Soviet space programme was built on an ambiguity of purpose. The R-7 started life as an intercontinental ballistic missile and it was for this purpose that it was valued by the Soviet political leadership, be that Stalin or his successor Nikita Khrushchev. Korolev and Tikhonravov, while prepared to build the R-7 for this purpose, were in reality interested in using the rocket to send satellites into space. With Stalin gone in 1953, it was possible to explore this possibility more openly.

Korolev and Tikhonravov formed a satellite team to design an Earth satellite in 1954. The project was approved by the Academy of Sciences and the government the following year. Various designs were explored and the original plan was to launch into orbit a large scientific satellite, object D, close to the payload limit, 1,350 kg. The announcement of an International Geophysical Year prompted consideration of the launch of an Earth satellite by the United States. Determined to beat the United States, the Soviet project was simplified in December 1956 and a smaller satellite decided on, called the PS (Preliminary Satellite). The real problem though was not the satellite, but getting the rocket flightworthy. The R-7 underwent static tests in Zagorsk that year, but failed three launch tests in summer 1957, eventually flying a suborbital payload to the Pacific Ocean in August 1957.

The Soviet satellite project was reported openly in the Soviet media, with full details being published of the first two missions some time in advance, with a crescendo of information as Russia celebrated the 100th anniversary of Tsiolkovsky's birth in September 1957. Apart from American intelligence, nobody seemed to treat the reports seriously in the West.

Sputnik entered orbit on 4th October 1957. Nothing was ever the same again. Most of the great historical events make an immediate impact, sometimes to fade over time. Sputnik was different. *Pravda* did report the launching the next day, but well

Mikhail Tikhonravov's satellite team

down the page, blandly headed 'Tass communiqué'. In the West, the British Broadcasting Corporation announced the launching at the end of its late news bulletin, a certain vocal hesitancy indicating that the station did not know exactly what to make of the event.

Whatever about the political leadership, ordinary people knew. As Sergei Korolev and his colleagues took the long train journey back from Baikonour cosmodrome, people stormed onto the platforms to stop the train, demanding to meet the engineers concerned. There was a palpable, rising air of excitement as they drew closer to Moscow. By this stage, people throughout the Soviet Union were talking excitedly about this extraordinary event. The next day, *Pravda* made the satellite the *only* front-page story. The American reaction was apoplectic. To bring home news of the event to its readers, the *New York Times* used a print face larger than any used since the day after the attack on Pearl Harbour. Congressional committees convened hurriedly to discuss the crisis in American science and technology. The rocket body of the Sputnik entered orbit and could be seen tracking across the cold night skies of North America, young and old going out into their back yards to watch the amazing spectacle. The long aerials on the back of the spacecraft transmitted *beep! beep! beep!* signals that could, quite deliberately, be picked up by relatively simple receivers. Khrushchev was soon bragging about Sputnik's achievements to foreign leaders, deriving double satisfaction from the fact that it had made an impact on his rivals in such an unexpected way.

THE IDEA OF AN INTERPLANETARY PROBE

Taking advantage of the public reaction to Sputnik, on 28th January 1958 Tikhon-ravov and Korolev sent a letter to the Central Committee of the Communist Party of the Soviet Union and the government called *On the launches of rockets to the moon* (sometimes also translated as *A programme for the investigation of the moon*), followed on 5th July 1958 by the more elaborate *Most promising works in the development of outer space* (note likewise that variations of this title have also appeared – e.g., *Preliminary considerations for the prospects of the mastery of outer space*, probably also a function of translation). This was an audacious plan outlining a vast pro-gramme of space exploration [1]. It included:

- Sending robotic spacecraft to Mars and Venus.
- Upgrading the R-7 launcher to a four-stage version to make this possible.
- Eventually, manned flights to the moon, Venus and Mars.
- Development of the critical path technologies to support these developments.

Specifically, the July 1958 memorandum proposed the exploration of Mars and Venus 'by automatic apparatus and returning to Earth's vicinity with photographic and other data' over 1959–61. Later, research should be carried out into the surface of the planets (1963–6). The critical path technologies identified were ion engines for interplanetary flight and the development of systems for long-range radio communications.

An important consideration was the Venus and Mars 'window'. Although all the inner planets of the solar system circled the Sun in the same direction at a similar angle to the ecliptic and in roughly circular orbits, the different speeds of these orbits took both planets in paths that were at times quite distant from Earth. Ideally, probes to the planets should be launched as their respective paths around the Sun drew closer as

Mikhail Tikhonravov

they passed one another, so that at interception Earth would be at its shortest distance to Venus within its solar orbit, or Mars farther out, so as to maximize the chances of receiving signals. The period most suitable for launching was called the 'Venus window' or the 'Mars window'. Generally, Venus windows took place every 18 months, with four to five month transit times. Mars windows took place every 25 months, with longer transit times of seven to eleven months. Due to the fact that the orbits of the planets around the Sun are not entirely circular, but slightly elliptical, some windows are better than others. Sometimes the Earth–Mars distance can be quite short, leading to quick transit times and/or the possibility of flying bigger payloads. In 2001, for example, the distance between Earth and Mars was only 40 million kilometres, the planet was a bright rusty object in the evening sky all that northern hemisphere summer and it was a good time for Mars missions. Other Mars windows have been less favourable, forcing designers to make trade-offs between payloads and transit times. The next windows were August 1958 (Mars) and June 1959 (Venus). Korolev wished to send spacecraft to both planets then.

No sooner had the first Sputniks been sent into orbit (Sputnik 2 followed in November 1957) than Korolev began to plan missions farther afield. The first lunar missions were planned for 1958. To do this, a small upper stage must be added to the R-7 rocket to send the spacecraft farther and faster [2]. Earth orbit required a velocity of about 7 km/sec, while lunar (or interplanetary) missions required more, a velocity of 11 km/sec.

Korolev asked two rocket engine designers to build him such an upper stage. He asked the man who had designed the engines for the R-7, called the RD-107 and RD-108, Valentin Glushko. Here, Glushko offered a powerful upper stage, called the RD-109; the new R-7 rocket would be given the design code of 8K73. He also asked another designer, Semyon Kosberg (1903–1965), who offered him an upper-stage engine called the RD-105. The code for this version of the R-7 would be the 8K72. As they did so, the preliminary designs were done of the Venus and Mars probes in early 1958 with a view to launching a 500 kg spaceship to Mars in August 1958 and Venus in June 1959. Calculations of the trajectories for the Mars and Venus windows were done by Mikhail Tikhonravov with the assistance of the leading applied mathematician of the Soviet Union and vice-president of the Academy of Sciences, Mstislav Keldysh.

It was soon realized in OKB-1 that the August 1958 Mars target was unrealistic. The 8K72 upper stage, using Kosberg's RD-105, was not ready until 2nd September 1958, when the first moon probe was launched. Kosberg's engine used oxygen and kerosene, had a thrust of 49 kN and a velocity of 3,099 m/sec. This failed, as did a successor on 12th October 1958. The launcher was grounded until the cause was investigated. The third launch failed on 4th December 1958.

News of some of these developments leaked out to the West. Until 1958, the Soviet space programme had been comparatively open, the Sputnik missions having been well signalled in advance and the scientific personalities of the programme identified in the press. Soviet plans for space exploration were widely discussed in the newspapers, with articles by the various designers. The launches of the first two Sputniks were preceded by a series of announcements about their forthcoming missions, orbits, frequencies and technical aspects, to such an extent that the launch

The calculator of trajectories – Mstislav Keldysh

ings should have taken no one by surprise. As an example of this, on 10th January 1959, Academician Anatoli Blagonravov expressed the hope that that June the Soviet Union would send a payload of up to 340 kg to Venus – which was indeed the intention at that time.

But this marked an end of anything but the broadest public statements about Soviet space intentions. During 1958, a mighty battle had waged within the Soviet political establishment as to how the space programme should be reported. That year, in April, the Soviet Union had attempted to launch the original satellite planned in 1956, the 1,300 kg object D . When the launching went astray, the decision was taken then *not* to announce failures (it was eventually launched in May). Apparently, the cult of scientific and industrial progress could admit no shortcomings, setbacks or failures. Over the years, though, Western experts were able to piece together the many launchings that went wrong [3].

Moreover, identity of the personalia of the Soviet space programme was now classified. Exceptions were made for one or two Soviet scientists already known in the West, principally the prominent personalities of the Academy of Sciences who were already known to international gatherings and who could not be easily 'disappeared'. But chief designer Korolev, chief engine designer Glushko and chief theoretician Mikhail Tikhonravov all retreated to the shadows, and the Soviet space programme was thereafter announced as 'the collective achievement of the Soviet people as a whole'. These two related decisions were slugged out between the chief ideologist of the Communist Party of the Soviet Union, Mikhail Suslov and Academy of Sciences Vice-President Mstislav Keldysh. Suslov won. For the record, a similar battle took place in the United States as to whether the man-in-space programme should be classified and militarized. Despite strong pressure to the contrary, President and General Dwight Eisenhower decided, to his everlasting credit, that it should be an open, civilian programme.

THE FIRST COSMIC SHIP

Eventually, the 8K72 launched the First Cosmic Ship (later known as Luna 1) on 2nd January 1959. It missed its target, the moon. The First Cosmic Ship did become, albeit inadvertently, the first Soviet spaceship to venture into interplanetary space and orbit the Sun. The First Cosmic Ship carried instruments for measuring radiation, magnetic fields and meteorites [4].

Instruments on the First Cosmic Ship
Gas component of interplanetary matter.
Magnetometer (fields of Earth and moon).
Cosmic radiation detector to measure:
  – meteoric particles and photons;
  – heavy nuclei in a primary cosmic radiation trap;
  – variations in cosmic ray intensity.
1 kg of sodium vapour.

Passing the moon at a distance of 5,995 km some 34 hours after leaving the ground, it went on into orbit around the Sun between the Earth and Mars, becoming the first Soviet interplanetary mission. *En route* it found that the moon had no magnetic field and that the Sun emitted strong flows of ionized plasma, marking the discovery of the 'solar wind'. Much later and for similar reasons, two more moon probes were to find their way into solar orbit and interplanetary space (Luna 4, 6). The United States' first spacecraft to enter solar orbit, Pioneer 4, also a moon probe, followed two months later. In March 1960, they followed with a dedicated deep space probe, Pioneer 5, specifically designed to operate in solar orbit.

Of the first four lunar attempts, three had failed completely, while the fourth had been inaccurate. This was not the failure of Kosberg, for his engine got the first chance to fire only with the First Cosmic Ship. The chances of a successful Venus mission in June 1959 receded. Glushko was still unable to get his more powerful RD-109 ready and it was eventually cancelled in 1959 without a prototype having been built.

PLANNING THE FIRST VENUS AND MARS MISSIONS

After a while, it was recognized that a better course of action would be to make less hurried preparations for the next, 1960 window to Mars and the 1961 window to

The First Cosmic Ship

Venus. A key factor, identified at an early stage, was the finding that the direct ascent to Mars and Venus meant a very tight launch window and imposed considerable constraints on payload. With direct ascent, the smallest error in the launch trajectory, even from early on, would be magnified. A change of plan was suggested by Korolev's designer, Gleb Maksimov. He calculated that putting the spacecraft into Earth parking orbit first provided much greater flexibility. In effect, the starting point for the mission could be any number of points in Earth parking orbit, not the single ground fixed location. Not only that, but parking orbit enabled the weight of the payload to be increased and more than doubled from the 200 kg originally anticipated. The downside was that this required putting a powerful stage in Earth orbit first. The engine would have to ignite after a period in Earth orbit. Although this seemed like a small detail, this requirement actually turned into a major technical problem area, for igniting propellants floating in weightlessness turned into a formidable challenge.

From early 1958 to August 1959, the Applied Mathematics Division of the Mathematical Institute of the Academy of Sciences carried out the calculations necessary for a flight to Mars in October 1960 and Venus in February 1961. They

Gleb Yuri Maksimov

found that the windows made possible a probe of 500 kg to Mars and 643 kg to Venus. The decision was taken, one which was continued into the 1980s, of always launching at least one spacecraft per launching period. It was calculated that the additional costs were only 15% to 20% more than that of a single launch and these economies of scale improved the chances of at least one positive outcome [5].

Gleb Maksimov also argued, successfully, for a standard spacecraft design to be adopted, one which could be multiply-produced but the instruments carried and the missions varied according to individual requirements. The series was given the title of 1MV (number 1 series for Mars and Venus). He also persuaded Korolev that interplanetary spacecraft should carry a mid-course correction engine to refine their trajectory and deploy a large parabolic antenna at planetary encounter to relay data back to Earth at this, the most important phase of the mission.

By happy coincidence, spaceships sent to both planets at this time would arrive at the same time, the third week of May 1961. The Mars probes would arrive on 13th and 15th May 1961. Proposals for a first set of missions were forwarded to the government by Sergei Korolev and approved in a resolution dated 10th December 1959: *On the development of space research*. An interdepartmental scientific and technical council was appointed to supervise the forthcoming missions, comprising Mstislav Keldysh (chairperson), Sergei Korolev, Anatoli Blagonravov and Konstantin Bushuyev (assistants to Keldysh) as well as Valentin Glushko, Mikhail Ryazansky, Nikolai Pilyugin, Mikhail Yangel, Georgi Tyulin and Vladimir Barmin.

The development plan for the 1MV missions was approved by Korolev on 28th February 1960, and this stipulated the timetable: the completion of design drawings (two weeks!), construction (April), integration (June), testing (August) and delivery to the launch site (September). On 15th March 1960, Vice-President of the Academy of Sciences Mstislav Keldysh approved a paper called *Designing spacecraft for Mars missions*, setting down the objective of photographing Mars from between 5,000 and 30,000 km and carrying instruments that might detect plant or animal life. Specific instrumentation would be:

- Cameras for the photography of Mars, with 750 mm cameras enabling the detection of surface details of 3 to 6 km, with picture sizes of 50 by 150 mm.
- Infrared instruments to detect plant life or other organic compounds on the surface of Mars.
- Ultraviolet instruments.

The theoretical calculations for the mission were recalculated and issued in April 1960 as *On predicting the accuracy of the trajectory of moving 1M spacecraft* and suggested the best possible launch date of 27th September 1960 [6].

THE ROCKET FOR THE MARS, VENUS PROBES

Gleb Maksimov had outlined how, replacing the 8K72 with a more powerful upper stage and using a parking orbit, larger payloads could now be sent to the planets. Construction of a new version of the R-7, called the 8K78, was approved by government resolution of 4th June 1960. It had to be got ready in fewer than three months, although some of its components had, thankfully, already been in development.

The 8K78 was a key development and became a cornerstone of the Soviet space programme, with its derivatives still flying almost 50 years later. The R-7 series as a whole, including the 8K78 offspring, became the most flown rocket in history, developed in several major versions and passed the 1,700 mark in 2005. The European Space Agency constructed a new launch base for the rocket in Kourou, French Guyana in 2007, giving this venerable rocket a new lease of life.

The following were the key elements of the 8K78:

- Improvements to the RD-108 block A and RD-107 block B, V, G and D stages of the R-7, with more thrust, higher rates of pressurization and larger tanks, developed by Glushko's OKB-453. These gave an additional several per cent more thrust.
- A new upper, third stage, the block I, developed with Kosberg's OKB-154, called the RD-0107 or 8D715K and based on the second stage of the R-9 missile. His engine used oxygen and kerosene, had a thrust of 67 kN and a velocity of 3,334 m/sec and a burn time of 200–207 sec.
- A new fourth stage, the block L, designed within OKB-1 by Vasili Mishin, also called the S1.5400 engine. Block L was 7.145 m long, the first Soviet rocket with a closed-stage thermodynamic cycle, with gimbal engines for pitch and yaw and two vernier engines for roll. It used liquid oxygen and kerosene.
- New guidance and control systems, the I-100 and BOZ.

In this new approach, the first three stages (blocks A, B–D, I) would put block L and the payload in Earth orbit. Block L would circle the Earth once in parking orbit before firing out of Earth orbit for the planets. The use of parking orbit with block L increased the payload from 200 kg to 1,000 kg (in practice, the payload for the first set of Mars and Venus probes would be less, in the order of 600 kg to 800 kg).

8K78

Guidance systems were developed both for the parking orbit and the interplanetary probes themselves by Mikhail Ryazansky (1909–1987), one of the council of chief designers going back to 1946.

Block L was designed to work only in a vacuum, coast in parking orbit and then fire planetward. A device called the BOZ (*Blok Obespecheyna Zapushka*) or Ignition Insurance System would guide the firing system. A new orientation system for blocks I and L, called the I-100, was devised by Scientific Research Institute NII-885 of Nikolai Pilyugin. Block L's S1.5400 engine was the first Soviet rocket designed to be used and re-used in a vacuum and had a specific impulse of 340 sec, the highest achieved by any rocket at that time. The engine used titanium alloys that enabled it to work at temperatures of 700°C. A first production run of 54 S1.400 engines was started in May 1960 and all passed their firing tests. A system called the Paused Stabilization and Orientation System was installed to ensure that block L would be at the right attitude for the trans-Mars and trans-Venus burn. This was a roll, pitch and yaw system, using small nozzles of 10 kg thrust. The block L project was led by a new engineer to the

OKB-1 team, Sergei Kryukov. He was a graduate of Moscow Higher Technical School and had, at the end of the war, been one of the team sent to Germany to scavenge for parts of the German A-4.

8K78 rocket for 1960–1 Mars, Venus missions

Total length	42 m
Diameter	10.3 m
Total weight	305 tonnes
of which frame	26.8 tonnes
propellant	279 tonnes
Burn time first stage (block A)	301 sec
Burn time second stage (block B)	118 sec
Burn time third stage (block I)	540 sec
Burn time fourth stage (block L)	63 sec

Block I was completed in May 1960. The fourth stage, block L, took longer. Block L was ordered in January 1959 and the blueprints approved in May. The first two stages, with block I, but without block L, were fired in suborbital missions into the Pacific Ocean on 20th and 30th January 1960. That summer, block L was first tested aboard Tupolev 104 aircraft, designed to simulate weightlessness. The complicated block L took a long time to certify as flight-ready and the ground tests were not complete until well into the Mars window, during the first week of October 1960.

TRACKING SYSTEM

The importance of a long-range communication system was stressed in the early Korolev/Tikhonravov papers. For Sputnik, the Soviet Union built a network of 13 visual and telemetry tracking stations, but these would not be adequate for deep space communications. A new system was specified, able to send and receive signals at a distance of 300m km.

In fact, progress had already been made on a long-range tracking system and the main site, Yevpatoria, had already been chosen in 1957 by Korolev himself. His chief colleague in this was Mikhail Ryazansky. Offering a southerly latitude, which was helpful for tracking the planets, Yevpatoria was in the Crimea and close to the Black Sea resorts. Originally the facility was named the MV (Mars Venus) Centre but was later called the TsDUC, or Centre for Long Range Space Communications. The first radio space-tracking facilities were built there in 1958 for the purpose of following the first lunar probes. There, a 22 m dish ground station on Kochka Mountain, near the Simeiz Astrophycial Observatory, had been operational since September 1958.

The TsDUC actually comprised two stations with two receivers (downlink) and one transmitter (uplink), facilitated by a microwave station. Both were close to the Crimean coast, west of Yevpatoria, which has its own airfield. The microwave station

Mikhail Ryazansky

transmitted data from the receiver stations to another microwave system in Simferopol and thence on to other locations in the USSR. The records are confusing as to what was actually built at the time and where and little was said about them, presumably to hide Soviet tracking capabilities from the snooping Americans. Although there are reports of a single dish being completed at this time, official details are sparse and tantalizing ('over 1,000 tonnes in weight, as tall as a 12-floor building'). We know that the Americans had intelligence maps of the Yevpatoria system from 1962 [7], but it would be surprising if they had not had good details a little earlier. Their illustrations of the complex were refined when pictures were taken by a KH-4 spy satellite overflight of June 1968.

In the event, construction at Yevpatoria went through three phases. The initial system was completed in 1960. Then new dishes called the Saturn system were added between 1963 and 1968 (32 m). In the third phase, a really big Kvant dish (between 79 m and 85 m) was completed in 1979, with a weight of 1,700 tonnes, making it the largest movable structure of its kind in the world.

For the moment, two sets of eight individual duralium receiving dishes of 15.8 m were built on a movable structure, designed to tilt and turn in unison. Two were built 600 m apart at what the Americans called 'north station' and a set of half the size, 8 m transmitting dishes called Pluton at what they called 'south station'. North station was for receiving signals, south station for sending them.

North station was the largest complex of the two, surrounded by 27 support buildings, 15 km west of Yevpatoria. To construct the receiving stations, Korolev was forced to improvise. He came up with the idea of using old naval parts for the station: a revolving turret from an old battleship, a railway bridge for support and the hull of a scrapped submarine. The dishes had to be sensitive. It was calculated that – when the first spaceship to Venus arrived at the evening star 112m km away – the strength of signal would be only 10^{-22} watt per $1\,m^2$ the Earth's surface. They received signals on the following frequencies: 183.6, 922.763, 928.429 MHz and 3.7 GHz.

South station was to the southeast and much closer to Yevpatoria, 9 km. It comprised one, later eight 8 m dishes in a similar configuration to, but half the size

Dish at Simferopol

of the duo at north station. Transmission power was rated at 120 kW and its range was estimated at 300m km. Transmissions were sent at 768.6 MHz. The Pluton transmitter was enormously powerful for its day, able to send a signal to a probe at Venus encounter at a strength of 15 watts.

Even though chief designer Yevgeni Gubenko died in the middle of construction, Yevpatoria station went online on 26th September 1960, six days after the start of the Mars window and a day before the optimal best day for launching, the 27th. The facilities there were originally quite primitive, ground controllers being provided with classroom-style desks, surrounded by walls of computer equipment. Modern wall displays did not come in until the mid-1970s. Despite this, it had the highest capacity of any tracking system in the world until NASA's Goldstone dish went into operation in 1966 [8]. To be precise, TsDUC comprised a western station (Yevpatoria) and an eastern station in Ussuriisk, near Vladivostok in the far east. Much less is known about the Ussuriisk station and the eastern end of the country was generally closed to visitors.

Until a mission control was opened in Moscow in 1974, Yevpatoria remained the main control for all Russian spaceflights, not just interplanetary ones. It was normal for the designers to fly from Baikonour cosmodrome straight to Yevpatoria to oversee missions. The Americans, by contrast, had a worldwide network of tracking stations, with large dishes in California, South Africa and Australia. Dependence on one station at Yevpatoria imposed two important limitations on Soviet interplanetary

The Pluton system

probes. First, the arrival of a probe at a distant planet had to be scheduled for a day and a time of day when the planet concerned was over the horizon and visible in Yevpatoria, so schedules had to be calculated with great precision many months in advance. Second, there was no point in having Soviet deep space probes transmit continuously, for their signals could not be picked up most of the time when Yevpatoria was out of sight. Instead, there would be short periods of concentrated transmission, called 'communications sessions', scheduled in advance for periods when the probes would be in line of sight with Yevpatoria. This required the use of timers and sophisticated systems of control, orientation and signalling.

Korolev and his colleagues attempted to get around the limits imposed by the Yevpatoria station. If they lacked friends and allies abroad to locate tracking dishes, there were always the oceans. Accordingly, three merchant ships were converted to provide tracking for the first Mars and Venus missions. They were the *Illchevsk*, *Krasnodar* and *Dolinsk* and their main role was to track the all-important blast out of parking orbit, which was expected to take place over the South Atlantic. These ships were a helpful addition, but they had limitations in turn. First, ships could not carry dishes as large as land-based dishes; and, second, they were liable to be disrupted in the event of bad weather at sea, which made it difficult to keep a lock on a spacecraft in a rolling sea. Later, in the mid-1960s, the Yevpatoria system was supplemented by a 32 m dish, called Saturn P-400. In addition to the new Saturn dish at Yevpatoria, five other Saturns were located in Baikonour, Sary Shagan (Balkash), Shelkovo (Moscow) and Yeniseiesk (Siberia).

Preparing for Mars and Venus: key dates

5 Jul 1958	Mikhail Tikhonravov and Sergei Korolev: *Most promising works in the development of outer space*
2 Jan 1959	First Cosmic Ship
20 Jan 1960	First suborbital test of block I into the Pacific Ocean
30 Jan 1960	Second suborbital test of block I into the Pacific ocean
15 Mar 1960	Mstislav Keldysh: *Designing spacecraft for Mars missions*
Apr 1960	Mstislav Keldysh: *On predicting the accuracy of the trajectory of moving 1M spacecraft*
Summer 1960	Tests of block L on Tupolev 104 jet aircraft
26 Sep 1960	Commissioning of Yevpatoria TsDUC

THE MAN-TO-MARS PROGRAMME IN THE 1950s

Even as the Soviet Union now prepared to launch its first spacecraft to Mars in September 1960, an ambitious enough undertaking in itself, a significant effort was under way behind the scenes to develop plans for a manned flight to Mars.

Extraordinary as it may seem now, Soviet plans for a manned flight to Mars date to 14th September 1956. Even though Korolev had yet to fly his first large rocket, the R-7, he was already thinking ahead to its much larger successor, to which he gave the relatively bland name of N-1, 'N' standing for *Nositel* or carrier, with the industry code of 11A51. First sketches of a rocket which could, among other things, fly cosmonauts to Mars appeared in OKB-1 on 14th September 1956. The concept was brought to the Council of Chief Designers on 15th July 1957, but it did not yet win endorsement. The N-1 at this stage was a large rocket able to put 50 tonnes into orbit.

The R-7 had been presented – and approved by the Soviet leadership – as a military intercontinental ballistic missile, later adapted to launch an Earth satellite. The N rocket, by contrast, was a universal rocket with broad applications. Korolev kept these purposes deliberately vague and, in order to keep military support for the project, hinted darkly at how the N-1 could launch military reconnaissance satellites.

In reality, a detailed read of the N-1 and its evolution suggests that Korolev intended to use the N-1 foremost as the basis for manned interplanetary exploration. The evidence may be found in an explanatory note sent to the Soviet government on 16th February 1959, signed by Sergei Korolev and Mstislav Keldysh, calling for the creation of heavy-lift rockets with the goal of launching manned interplanetary spacecraft; and a draft decree sent to the Kremlin 12th April 1960 by Sergei Korolev outlining future goals of the Soviet space programme, including an interplanetary manned spacecraft with a crew of two or three to fly by and land on Mars or Venus. The spacecraft would weigh 10 to 30 tonnes, including 3 to 8 tonnes for payload. For flyby expeditions, unmanned probes would be dropped. Three or four spacecraft would fly in formation, one serving as a backup return craft. The document was rewritten and re-sent to the government on 30th May 1960 and the spaceship redesignated Object KMV (*Korabl Mars Venus*), with development planned over 1962–5. These missions fitted perfectly into the profile of the N-1 [9].

The N-1 languished for several years. Unlike the R-7, the project did not have any precise military application and as a result the military would not back it. The situation changed on 23rd June 1960 when the N-1 was formally approved by Resolution # 715-296 of the government and party called *On the creation of powerful carrier rockets, satellites, space ships and the mastery of cosmic space 1960–7*, a substantial plan for long-range space exploration. Encouraged by the success of early Soviet space exploration, aware of reports of the developments by the United States of the Saturn launch vehicle, the Soviet government issued a party and government decree which authorized the development of large rocket systems, such as the N-1, able to lift 50 tonnes. The decree also authorized the development of liquid hydrogen, ion, plasma and atomic rockets. The 1960 resolution included approval for a N-2 rocket (industry code 11A52), able to lift 75 tonnes, but it was dependent on the prior development of these liquid hydrogen, ion, plasma and nuclear engines, suggesting it was a more distant prospect. The 1960 resolution also proposed circumlunar and circumplanetary missions. The implicit objective of the N-1 was to make possible a manned mission to fly to and return from Mars.

KORABL MARS VENERA AND THE TIZHULY MEZHPLANETNY KORABL (TMK)

Specifically, the 1960 resolution included project KMV, or *Korabl Mars Venera*, or Mars Venus spaceship, for circumplanetary missions to Mars and Venus to be developed over 1962–5. Such a mission was mapped out by a group of engineers led by Gleb Yuri Maksimov in OKB-1's Department # 9, overseen by Mikhail Tikhonravov, both members of the original satellite team of 1954–5. They were joined by two talented designers and both subsequent cosmonauts, Konstantin P. Feoktistov and Valeri Kubasov. Gleb Maksimov and Konstantin Feoktistov were born the same year, 1926 (Maksimov died in 2000, Feoktistov is still alive). This would not be a landing, but a circumplanetary mission of between one and three years in duration, depending on the trajectory flown. Government resolution or not, such studies had already commenced in 1956 under the direction of Mikhail Tikhonravov, the initial one being called the MPK or the Mars Piloted Complex [10].

Maksimov's studies postulated a heavy interplanetary ship, or TMK (*Tizhuly Mezhplanetny Korabl*), which would fly past Mars and return within a year. The TMK required the assembly in Earth orbit of a 50-tonne Mars spaceship. The crew would be brought up separately by an R-7 rocket to board the station in Earth orbit. The spaceship to actually fly past Mars, the TMK, with its crew of two to three cosmonauts, was not that different from the subsequent *Salyut* orbital stations, 20 m long, 4 m diameter, but with three 'floors' in the main module. There were solar panels, the spaceship would rotate around its long axis for gravity and there would be a biosphere called the SoZh for food and here algae would convert carbon dioxide into oxygen, food would be grown and human waste would be disposed. The 4 m (or, to be more

Konstantin Feoktistov

precise 4.1 m) diameter was a recurrent dimension of the Soviet space programme, for it was the maximum width that could be ferried by rail from Moscow to Baikonour.

The TMK had an instrument module, which included a radiation shelter and a centrifuge to create artificial gravity by rotation. The habitable sections had a diameter of 6 m. There would be a nuclear reactor able to generate 7 MW. This TMK-1 design was signed off on 12th October 1961 and a flight plan was even worked out, with departure from Earth set for 8th June 1971. Robots would be dropped off as TMK-1 passed over the red planet. A variation on the plan was called *Mavr*, a linguistic combination of Mars and Venera, involving a flyby of Venus on the return journey.

Even within Department #9 of OKB-1, there were rivalries. The first TMK-1 design was done by Gleb Maksimov for a *circum-Martian* mission. No sooner had Maksimov presented his first design on 20th April 1960 than Feoktistov persuaded Korolev that he should be allowed to go ahead with a more ambitious study for a manned *landing* on Mars. Korolev, remembering his friendship with the long-dead Friedrich Tsander, was delighted at the interest in Mars and spurred both teams on to work ever harder. Konstantin P. Feoktistov had a remarkable background. Born in Voronezh in 1926, he was a child prodigy who learnt Tsiolkovsky's formula at the age of ten, going on to master maths and physics. He was a scout during the war, captured by the Germans, shot as a spy and left for dead. Invalided for his bullet wound to the head, he entered the Bauman Technical College in 1943 and took his engineering degree in 1949, becoming a space designer in the years that followed. His character was evident in other ways: in a country where membership of the Communist Party was a test of loyalty and often necessary for promotion, Feoktistov never joined.

Feoktistov's team foresaw the assembly, in Earth orbit, of a manned spacecraft for a crew of six, using ion electrical engines propelled by nuclear power. This was a much more ambitious enterprise and two N-1s would be involved to assemble a much

Korolev and Feoktistov

larger structure. Five movable platforms would be landed on Mars: one to drill soil; a second to launch aircraft; two with return rockets for the journey back to Earth; and a nuclear-powered generator to supply energy for the expedition. This would act as a train engine to pull the other platforms across the planet for a year. A base spacecraft would orbit Mars. This TMK was like a daisy stem (nuclear power plant at one end, crew quarters at the opposite). Their TMK-2 (the term TMK-E has also been used) was 123 m long, 19.6 m diameter, weighed 75 tonnes and had a crew of 10. It would use 7.5 kg low-thrust plasma engines to spiral out of Earth orbit.

According to Korolev's biographer, Jim Harford, the N-1 TMK study team (or teams) was typical of Korolev. Even before the first cosmonaut had flown, he had assembled a strong design team for a manned Mars mission. Korolev was always thinking not just one, but two or three steps ahead. He expected the same vision, passion, enthusiasm and round-the-clock effort from others that he showed himself – and he got it. The TMK Mars studies were the first serious engineering studies of a manned mission to the planet ever undertaken. They set the N-1 design at 50 and then 75 tonnes, that which was necessary for assembling a manned Mars spacecraft.

This was not just a paper study. On 3rd May 1961, OKB-1 agreed the design of the TOS or Heavy Orbital Station (*Tizhuly Orbitaly Stantsiya*), which would essentially be the design of the TMK. He proposed it be flown in Earth orbit for a year, thereby debugging the TMK for the subsequent Mars flight. The existence of the TOS was not known about in the West until the 1980s, when Korolev's colleague Mstislav Keldysh published all the old designs in *The creative legacy of Sergei Korolev* (1980). Here were found details of the TOS and a model TOS, shaped like a squat cylinder, was even assembled in OKB-1. Here too can be found notes by Korolev on the Maksimov and Feoktistov designs handwritten in September 1962. A conference on the development of biospheres for the Mars flight was held on 22nd July 1963 and attended by Mstislav Keldysh, Sergei Korolev and cosmonaut squad commander, Nikolai Kamanin.

The Mars designs occupied considerable time and effort in OKB-1 over 1959–63, all this at a time when NASA was putting the Apollo project together. The historian Assif Siddiqi [11] points out in his narrative that the TMK programmes were very real, tied down significant resources and were the primary objective of the N-1 programme. Had the Soviet space programme not been re-directed in August 1964 to respond to the American flight to the Moon, then the Soviet Union would undoubtedly have sent the first cosmonauts to fly past Mars in the 1960s. The N-1 design, although it proved problematical for a man-on-the-moon project, was actually perfect for the assembly of a Mars expedition in Earth orbit. Ambitious though the Apollo project was, Korolev had all along planned to go much farther. The slogan of the GIRD group, written by Friedrich Tsander, never referred to the moon at all. Instead, its motto was 'Onward to Mars!'

ANOTHER WAY TO GO: *KOSMOPLAN*

Improbable though it may seem now, TMK was one of two plans that existed at this time for the manned conquest of Mars.

Korolev had a formidable rival for the leadership of the Soviet conquest of the cosmos, a man almost unknown outside his own country: Vladimir Chelomei. His bureau, OKB-52, had built its reputation on missiles and – in the course of the 1960s – was later to develop a range of military space programmes. His much more blatant pitch for military space projects meant that he found it easier to command resources than Korolev, who was primarily interested in space exploration for its own sake.

Vadimir Chelomei was slightly younger than Korolev, born 30th June 1914 in Sedletse and later attending Kiev Aviation Institute. While an intern at the aviation plant in Zaporozhe in 1937, he proposed an untraditional solution to mechanical failures in plane engines, marking the start of a lifetime of untraditional approaches to engineering problems. He obtained a coveted 'Stalin doctorate' in 1940 and entered

Vladimir Chelomei

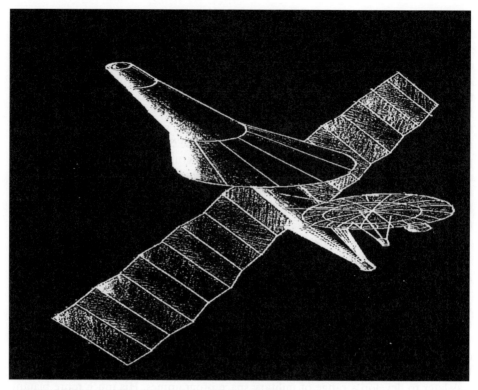

Kosmoplan

Baranov Central Institute of Aviation Building in Moscow in 1941. He attracted national attention in 1944 when he managed to reverse-engineer the German V-1 pulse engine cruise missile. From then on, he specialized in cruise, winged and military missiles, becoming chief of his own design bureau, OKB-52 in Reutov in 1955. By 1960, he felt confident enough to propose space rockets and rival Korolev and his deputy Vasili Mishin for leadership of the Russian moon programme.

Much to Korolev's annoyance, Chelomei attempted a 'grand entrance' to the space industry by proposing a series of projects to the political leadership in April 1960, the very time when Korolev was organizing his own plans. Vladimir Chelomei made a number of presentations of his projects to the Defence Ministry, the Military Industrial Commission, even Khrushchev himself and following a formal review they were approved on 23rd June 1960 by resolution of the party and government. The approved projects included an electric interplanetary spacecraft (*Kosmoplan*), a space-plane (*Raketoplan*) and satellite interceptor.

The one of most relevance here is the *Kosmoplan*, a solar electric spaceship designed to fly to Mars. Chelomei's designs were original, even idiosyncratic and would now attract the epithet of 'out-of-the-box thinking'. *Kosmoplan*'s structure looked like a toy glider, with two large, outstretched solar wings. Launched first into Earth orbit, it would generate low thrust from plasma engines to spiral gradually outwards from Earth orbit and then fly to Mars, using the same technique to return. Solar power and nuclear power were variously considered as energy sources, though the plasma propulsion engine was the same. At the top was a conical spaceplane. On reentry to Earth, it would extend a large umbrella to cushion its return to the Earth's atmosphere before landing like a plane.

The spaceship consisted of a long frame of two poles, with an engine at the rear and two cylinders in the middle, the second one rotating within the frame for purposes of artificial gravity. The reason the project did not progress further was the huge expense involved in developing the plasma engines. Chelomei deliberately used the term *plan* to distinguish his designs from the *korabl* or 'spaceship' of Korolev, whose approach he regarded as pedestrian. In reality, many of Chelomei's ideas were far ahead of his time. The spiral technique was eventually proven, putting Europe's small moon probe, SMART-1, into lunar orbit, but not until 2003. The umbrella technique of atmospheric reentry was eventually tested by Russia using an inflatable rubberized cone, but not till 2000.

Chelomei's plan was, like Korolev's Object KMV, approved in 1960. His design bureau, OKB-52, proceeded to sketch further details of *Kosmoplan*. It seems that the designers probably became aware of the practical difficulties fairly quickly and their efforts switched to projects that could be realized more speedily. A decree made by the government and party on 13th May 1961, *On the revision of plans for space objects toward accomplishment of goals of a defence nature*, suspended the *Kosmoplan* but switched resources from Korolev's OKB to Chelomei's so that the latter could concentrate on military space projects. Chelomei took the additional resources to expand his space empire and, decree or not, design work continued on the *Kosmoplan* for at least another four years. The battle to build a manned spaceship to Mars between Chelomei's OKB-52 and Korolev's OKB-1 was to resume when the manned programme to Mars recommenced in 1969 as the *Aelita* project (see Chapter 8, *Returning to the planets?*).

READY FOR MARS AND VENUS

Thus, by September 1960, Russia was nearly ready to send it first probes to Mars. Chief designer Sergei Korolev had moved quickly after the first flight of the R-7 and Sputnik to persuade the government that he should now go ahead with the first interplanetary spacecraft. The necessary upper stage could not be made ready in time for the first Mars window of 1958, nor the first Venus window of 1959, but the breathing space gave time for the preparation of a more powerful version of the R-7, the 8K78, with new upper stages, block I and block L. The infrastructure of tracking stations and ships had been put in place.

Even as the Soviet Union prepared to send its first Mars probes, work was well advanced on preparing a manned flight to Mars, Object KMV, later the TMK. The concept, using the N-1 rocket, had first been committed to paper in September 1956 and since then one, then a second design team had drawn up concepts for a manned Martian expedition, a flyby and a landing. The two TMK designs by Gleb Maksimov and Konstantin Feoktistov proceeded alongside development of the N-1 launcher and the first work on the TOS heavy orbital station. An innovative rival design, *Kosmoplan*, was prepared by the ambitious Vladimir Chelomei.

REFERENCES

[1] Siddiqi, Assif: *The challenge to Apollo*. NASA, Washington DC, 2000; Harford, Jim: *Korolev*. John Wiley & Sons, New York, 1996.
[2] For an account of the development of the R-7 for interplanetary missions, see: Varfolomeyev, Timothy: Soviet rocketry that conquered space. *Spaceflight*, in 13 parts:
   1 Vol. 37, #8 August 1995;
   2 Vol. 38, #2, February 1996;
   3 Vol. 38, #6, June 1996;
   4 Vol. 40, #1, January 1998;
   5 Vol. 40, #3, March 1998;
   6 Vol. 40 #5, May 1998;
   7 Vol. 40, #9, September 1998;
   8 Vol. 40, #12, December 1998;
   9 Vol. 41, #5, May 1999;
  10 Vol. 42, #4, April 2000;
  11 Vol. 42, #10, October 2000;
  12 Vol. 43, #1, January 2001;
  13 Vol. 43, #4, April 2001 (referred to as Varfolomeyev, 1995–2001).
[3] Clark, Phillip S.: Launch failures on the Soviet Union's space probe programme. *Spaceflight*, vol. 19, #7–8, July–August 1977.
[4] Don P. Mitchell has carried out significant research into the scientific instrumentation carried by Soviet interplanetary probes, as well as the allied areas of propulsion, cameras and telemetry. See: Mitchell, Don P. (2003–4):
  – Soviet interplanetary propulsion systems;
  – Inventing the interplanetary probe;
  – Soviet space cameras;
  – Soviet telemetry systems;
  – Remote scientific sensors;
  – Biographies, at *http://www.mentallandscape.com*
[5] Huntress, W.T., Moroz, V.I. and Shevalev, I.L.: Lunar and robotic exploration missions in the 20th century. *Space Science Review*, vol. 107, 2003.
[6] Varfolomeyev, Timothy:
  – The Soviet Venus programme. *Spaceflight*, vol. 35, #2, February 1993;
  – The Soviet Mars programme. *Spaceflight*, vol. 35, #7, July 1993.
[7] Grahn, Sven: Yevpatoria – as the US saw it in the 60s. Posting by Sven Grahn at *http://www.svengrahn.ppe.se* (2005).
[8] For an account of Soviet tracking systems, see Mitchell, Don P.: Soviet telemetry systems at *http://www.mentallandscape.com*
[9] Zak, Anatoli: Martian expedition, *http://www.russianspaceweb.com*
[10] Wade, Mark: *TMK-1*. Encyclopedia Astronautica at *http://www.astronautix.com*, 2005.
[11] Siddiqi, Assif: *The challenge to Apollo*. NASA, Washington DC, 2000; Harford, Jim: *Korolev*. John Wiley & Sons, New York, 1996.

3

The first Mars, Venus probes

Would they do all this, if they realize what really lies ahead? The first inevitable failures will discourage the faint-hearted and undermine the confidence of the public.

– Konstantin Tsiolkovsky, 1929

1M AND 1V SERIES, 1960

Chapter 2 described the new 8K78 rocket that would be used to send the first spacecraft to Mars and then Venus. What of the spacecraft themselves?

The 1960–1 Mars and Venus probes were called the 1MV series. These first Mars probes would be called the 1M and the first Venus probes 1V. The chief designer was Sergei Korolev and science leader was the vice-president of the Academy of Sciences, Mstislav Keldysh. It may be worth saying a little more about Mstislav Keldysh. He had a distinguished background, for his father was Vsevolod M. Keldysh (1878–1965), designer of the Moscow Canal, the Moscow Metro and the Dnepr Aluminium Plant. Mstislav was the Soviet Union's leading mathematician and by this time had become vice-president of the Academy of Sciences, to the outside world its guiding force. He had joined the Zhukovsky Central Institute for Aerohydrodynamics, TsAGI, as a young mathematician in the 1920s and quickly made an impression by combining mathematics with engineering to solve practical problems in aircraft design, such as propellor flutter and landing gear failure, winning a state prize in 1942. Very much an all-rounder, he branched into physics, being awarded his doctorate in 1938. He ran a lecture course in Moscow University from 1938 to 1958, focusing on complex variables, partial differential equations and functional analysis, being considered the father of the theory of function approximation in the complex domain.

Mstislav Keldysh

Once Stalin died and the Soviet Union resumed its development of computers, it was Keldysh who led Russia's rapid industrialized computerization at the Institute of Applied Mathematics. In recognition of his achievements, he became president of the Academy of Sciences in 1961. Not one to treat the position as a sinecure, he used his position to reinforce Soviet studies of basic science, while at the same time pushing back the frontiers of new sciences, such as quantum electronics, holography, genetics and molecular biology [1]. Even when Suslov had consigned the chief designer and his associates to anonymity, Keldysh was permitted to continue as the public face of Soviet science, speaking to journalists in Moscow and going to many international conferences abroad, visiting Britain and the United States. The greying Keldysh was an important interlocutor between the design bureaux on the one hand and the party and government on the other, Keldysh often being called in to adjudge the merits of the many competing projects put up by the design bureaux.

1MV design team

Chief designer:	Sergei Korolev
Science leader:	Mstislav Keldysh
Designer:	Gleb Yuri Maksimov
Flight programme and spacecraft logic:	A.G. Trubnikov
Development and design:	L.I. Dulnev
Ballistics:	Valeri Kubasov

Korolev aimed to send three spacecraft to Mars in 1960, the third to land there. For the first two, the objective was to fly past and photograph Mars, while also carrying out experiments in interplanetary space on the way to the planet. Planning a landing on Mars during the first set of missions there was daring. The original 1MV plan had been to detach and drop a small lander the size of a television set onto the two respective planets. During the summer of 1960, landing tests of a lander had been carried out using the R-11A suborbital rocket. Mockup landers were flown to an altitude of 50 km to test the parachute systems of the lander, which would weigh up to 285 kg. Large parachutes were tested for the Mars landers and smaller ones for Venus, reflecting the respective thinness and thickness of their atmospheres. In the event, only two Mars probes were completed, both the flyby probes. Work on the third probe, the proposed lander, did not even get far into the design stage.

The 1960 Mars probes each weighed 640 kg and carried eight instruments, weighing 10 kg. *En route* to the planet, a decimetre wavelength radio transmitter was used. For solar power, there were $2 \, m^2$ solar panels supplying a zinc battery. The cameras, designed by Petr Bratslavets, were the same as those developed for the Automatic Interplanetary Station which circled the farside of the moon in October 1959. Pictures of the encounter with Mars would be transmitted by a 2.33 m high-gain antenna on the 8 cm wavelength. A series of scientific instruments was planned.

Petr Bratslavets

1M series, 1960 instruments and their designers

Magnetometer	Shmaia Dolginov
Ion traps to measure solar plasma	Konstantin Gringauz
Cosmic ray detectors	Sergei Vernov
Micrometeorite sensors	Tatiana Nazarova
Radiometer	
Charged particle detector	
Spectroreflectometer to detect organic life	
Infrared and ultraviolet spectrometers (removed)	Alexander Lebedinsky
Camera system (planned, not carried)	Petr Bratslavets

The magnetometer dated originally to a 1956 meeting between chief designer Sergei Korolev and the first head of the space Magnetic Research Laboratory, Shmaia Dolginov (1917–2001). He headed the laboratory in the Institute of Terrestrial Magnetism (IZMIRAN) where he had mapped the Earth's magnetic field by sailing around the world in wooden ships using no metallic, magnetic parts. He worked with Korolev to install a magnetometer on Sputnik 3, which duly mapped parts of the Earth's magnetic field. Now it would map magnetic fields on the way to and at Mars. Ion traps were used to detect and measure solar wind and solar plasma and were developed by Konstantin Gringauz (1918–1993), who had been flying his traps on sounding rockets as far back as the 1940s. He had famously built the transmitter on Sputnik and was the last man to hold it before it was put in its carrier rocket. The meteoroid detector was developed by Tatiana Nazarova of the Vernadsky Institute. Essentially, it comprised a metal plate on springs which recorded any impact, however tiny. The cosmic ray detector was developed by Sergei Vernov (1910–82) of the Institute of Nuclear Physics in Moscow, who had been flying cosmic ray detectors on balloons since the 1930s.

Whether they would all fly now came into doubt. The lander was not the only part of the programme to fall behind in the ambitious schedule laid down the previous February. The launch window for Mars opened on 20th September, with the following week being the best for a Mars launching, the 27th best of all. On the day the window opened, the radio system was giving trouble, had failed ground-testing and had still not left its production factory in Moscow. When it eventually arrived, duly passed and certified, it had to be integrated with the rest of the spacecraft. This time, there were further problems and it took two days to get the radio system to function with the television system.

By this time, there was no way the probe could be delivered to Baikonour and launched before 25th September. Every day that passed reduced the mass of the payload that could be launched. Accordingly, the television system was removed both to save weight and because of doubts that it would even work once it reached Mars. The ultraviolet spectrograph and infrared spectrometer were next to go. The loss of the latter was a big disappointment, because it was intended to analyze the darker regions of the planet and determine if there was vegetation or not. The 1MV had a mid-course correction engine: the KDU-414 engine, which used nitric acid and

Shmaia Dolginov

dimethyl hydrazine able to generate 200 kg of thrust. This would perform at least one course correction and would then be jettisoned in March 1961.

The first probe was delivered to Baikonour on 8th October, but without pressure-testing. The tracking system at least was ready, having been certified on 26th September. So too were the ships to monitor the critical blast out of parking orbit. The *Illchevsk*, *Krasnodar* and *Dolinsk* had left their Black Sea ports in August and had now been on station for more than two weeks. From this point on, the position of these tracking ships became a key indicator to Western intelligence of Soviet plans to send spaceships to the planets. All the ships had to pass through the narrow straights of

Konstantin Gringauz

Istanbul (Constantinople) where they could be spotted (their tracking aerials were a give-away) and then followed. Even in the 21st century, the role of tracking ships was to become important in watching the Chinese space programme, for they used a similar fleet for their spaceship, the *Shenzhou*. Western intelligence knew they must be waiting for something.

The first Mars mission, stripped of many of its components to save weight, launched from Baikonour on 10th October. During the firing of the second stage, vibration was so strong that it caused the gyroscope's pitch control system to fail. By the time the third stage was firing, the gyrocompass was able to detect that the rocket had gone far off course, far beyond the 7° permitted. There was no system for the destruction of the rocket then; the procedure adopted was to stop the engine firing, and this was done at 309.9 sec (in Russian, AVD, or *Avarinoye Vyklyuchennie Dvigatelei*, or emergency cut-off of engines). The rocket reached 120 km and then crashed back to Earth over eastern parts of Siberia.

When the second probe was launched on 14th October, at 290 sec into the mission the third-stage block I engine just failed to ignite. The block I and L upper stages, with their precious payload, dived back into the atmosphere where they broke up. An investigation found that the rocket had been doomed all along. A valve had failed and the kerosene had frozen when it was still on the launchpad [2] and was still frozen when it was supposed to flow into the turbopump. The disappointed *Illchevsk*, *Krasnodar* and *Dolinsk* returned to port, arriving back in the Black Sea in November. No one was the wiser as to whether the troublesome block L or the hastily assembled Mars probes would have worked.

Launcher for the first Mars probes, the 8K78

It so happened that during all of this Nikita Khrushchev was on a lengthy tour of the United States, having arrived on the 19th. Khrushchev loved to impress the Americans, and on his first visit there the previous year he had arrived splendidly, glowing in the Soviet Union's latest aerial triumph, the gleaming Tupolev 114 airliner, which, although it was propellor-driven, had astonishing range, speed and even two floors. Conventional wisdom has it that he timed his visit in order to be able to announce, in New York, Russia's latest triumph in space [3]. It would probably have been nice to have done so there. An analysis of Khrushchev's movements suggests that the speaking schedule of the United Nations was actually the determining factor, for his famous shoe-banging speech was on 12th October and he left the United States on the 13th. He would have been *en route* on his way back when the second Mars probe was launched on the 14th [4]. According to a defecting sailor, Khrushchev's ship included a model of the Mars probe which he had hoped to display, but it must have sailed back with him.

These launches were not announced at the time. Under international law, the space powers had agreed to announce all spacecraft arriving in orbit and give details of their trajectories (apogee, perigee, inclination and period). Because neither actually reached orbit, there was no legal obligation to report. The United States eventually published, in June 1963, a long and accurate list of Russian Mars, Venus and moon failures dating back to 1960 (it later transpired there were even more moon failures, going back to 1958). Some people assumed the Americans were making the lists up as some form of black propaganda, for no country could afford so many failures and still keep on trying. The Soviet government, for its part, snottily dismissed the allegations by saying that 'it is not important whether or not there have been failures when Soviet science and techniques are doing so wonderfully and are acknowledged throughout the world.' The Americans had installed electronic tracking stations along the Soviet Union's southern borders and they were able to pick up the radio traffic whenever launchings were due. U-2 spyplanes were put up from bases in Pakistan and Turkey to pick up electronic traffic. Some even flew over Baikonour cosmodrome and one even managed to spot an ascending rocket head for orbit. The 1960 Mars failures were probably high enough over the horizons of the American radars and their blips and signals would have been detected. The times of launch windows for Venus and Mars were well known to the Americans who were also able to identify the type of rocket being launched from its radar, radio and chemical signature. The movements of the tracking ships were a further confirmation. It was not difficult to guess that the unannounced events of 10th and 14th October were interplanetary attempts.

HIDING IN PLAIN VIEW: *TYZHULI SPUTNIK*

The first Mars probes shared a common design with what were intended as the first Venus probes. Even before the delayed Mars launches, Sergei Korolev had turned his attention in September 1960 to how the Mars design should be adapted for Venus, for which the forthcoming launch window was from 15th January to 15th February 1961. The probe was labelled the 1V.

Had all the four planned 1MV missions worked, the world would have witnessed a spectacular interplanetary *tour de force* as they arrived at their destinations in rapid succession. Arrival dates at the two planets were scheduled as follows:

Arrival dates of 1MV systems (1961)
11 May   First Venus probe
13 May   First Mars probe
15 May   Second Mars probe
19 May   Second Venus probe

The 1V series was originally designed as a lander. Here the overstretched Korolev ran out of time again and he never concluded the planned landing tests or design. It is possible that Korolev, being realistic, knew he would be lucky if the probe ever got that far, but much would be learned in the meantime, he hoped. The design was settled by Korolev on 1st January 1960. Because it was a re-design of the original 1V plan, it received a new designator, the 1VA. The dome was a thermal cover to protect the spacecraft during descent and buoyant so as to help the probe to float once it splashed down onto Venus' oceans [5]. Inside the dome was a pressurized globe float, carrying a pennant of the USSR with the Soviet coat of arms, an Earth globe and an inscription of the solar system, all designed to bob on the oceans of Venus. The dome included what was called a phase state sensor, a little like a builder's spirit level, designed to tell

AIS design

whether the spacecraft was stable or, if it moved, floating. The dome was expected to come free at some stage during the descent, but no diagram has ever been published to explain how.

The first 1V Venus probe launched on 4th February 1961. It did marginally better than the two Mars probes, actually reaching an Earth parking orbit of 223 km by 328 km. When the time came for the probe to leave Earth orbit, the timer failed to give the command for fourth-stage block L ignition. This was in turn due to a failure of the PT-200 transformer to supply current, which meant that the timer had not been powered up and so the signal was not sent.

This left the Russians with a problem as to how to explain the arrival of the satellite in orbit. Tracked by American antennae worldwide and with an international obligation to report on satellite launches, it could hardly be concealed. Because the Venus probe with the fourth stage weighed a record 6,843 kg, they ingeniously announced that it was a test of a 'heavy satellite' (*Tyzhuli Sputnik*) that had completed its mission on its first orbit. Western observers were unconvinced. Russia was preparing its first manned flight into space at that time and speculation inevitably arose that it was a manned spaceflight that had gone horribly wrong. Imaginative radio listeners in Italy even picked up the transmissions of heartbeats from the doomed cosmonaut allegedly on board – indeed, in the course of four years of listening to manned Soviet space failures the little Italian station run by the Cordolia brothers wiped out an entire squad of fictitious cosmonauts.

The *Tyzhuli Sputnik* decayed from orbit on 26th February, and some Western records, not entirely accurately, give it the designator of 'Sputnik 7'. Two years later, a young boy found parts of *Tyzhuli Sputnik* in Siberia's River Baryusha where they crashed back, including charred and bent pennants celebrating the first prospective landing on Venus. They had a circuitous subsequent history. The patriotic youngster brought them to the KGB which, when the Soviet Union dissolved, gave them to the Russian Academy of Sciences. To raise money for Russia's crumbling science programmes, they were soon then sold off at auction in 1996 in New York [6].

THE FIRST AUTOMATIC INTERPLANETARY STATION TO VENUS

At its fourth attempt to launch a planetary probe, the Soviet Union successfully dispatched what it called an Automatic Interplanetary Station to Venus on 12th February 1961. The spaceship first entered its parking orbit of 227 km by 285 km, 65.7° and at the end of its first parking orbit the fourth stage fired perfectly to send the station on a curving trajectory to Venus, due to arrive on 19th May. Weight was 643 kg and it was transmitting at 922.8 megacycles a second.

The Russians called the probe an 'Automatic Interplanetary Station' and it acquired the name of Venera 1 only later. This terminology was confusing, for the spacecraft that had circled the moon in October 1959 was also called the 'Automatic Interplanetary Station', though it was eventually, many years later, retrospectively named Luna 3.

This automatic interplanetary station was 2.035 m high, 1.050 m diameter, weighed 644 kg with a domed top, cylindrical body and two solar panels to soak up sunlight and turn it into electricity. It had instruments to send back information on radiation, micrometeorites and charged particles, a similar package to the 1M series. Cameras had been planned, but once again they were taken off to save weight. Thermal shutters opened and closed the hermetically sealed instruments from the heat and cold of deep space and aimed to maintain a temperature of 30°C. The solar panels had an area of $2\,m^2$. The dome was pressurized at 1.2 atmospheres. A mid-course correction engine was carried, although it was not identified in the often detailed explanatory diagrams of the station published as the time.

Automatic Interplanetary Station in-flight experiments, with their designers

Cosmic ray and gamma field detector	Sergei Vernov
Counter to measure charged particles of interplanetary gas and corpuscular streams from the sun	Konstantin Gringauz
Micometeoroid detector	Tatiana Nazarova
Magnetometer	Shmaia Dolginov

An orientation system obliged the solar panels to face the sun and recharge the chemical batteries whenever they discharged their energy. This was developed by Boris Raushenbakh (1915–2001) who had achieved fame as the designer of the orientation system used for the spacecraft that circled the moon in 1959, subsequently named Luna 3. Boris Raushenbakh was, as one can gather from his name, a German by background and was interned during the war, but built up an expertise in systems of orientation, sensors and the use of microjets in the Keldysh Institute.

Boris Raushenbakh

The station was expected to transmit every five days, first of all giving a location signal for 17 min and then a data stream, after which ground commands might be sent up – the 'communications session'. There were three antennae:

- A 2 m umbrella paraboloid made of fine copper mesh, to be lowered at Venus approach, transmitting on 8 cm and or 32 cm bands.
- A 2.4 m long omni-directional antenna for early stages of the mission at a wavelength of 1.6 m.
- A T-shaped antenna deployed on leaving the Earth's gravitational field until the approach to Venus, transmitting on 922.8 MHz at 1 bit/sec.

The telemetry system was designed and built by the Ryazansky design bureau. Up commands were sent on 770 MHz and 1.6 bit/sec. The spacecraft was designed to acknowledge commands sent and confirm execution.

Solar and star sensors, built by the Geofizika design bureau, would search for the sun and stars and once locked, compressed nitrogen jets would fire to orientate the spacecraft correctly. Position would be maintained by gyroscopes built by the Khrustalev bureau. It was intended that the probe be normally orientated toward the Sun so as to maximize electric power, but that during a communications session the Earth sensor would identify the blue planet, lock on and permit transmission.

The AIS on its way

The Automatic Interplanetary Station headed into regions of space never explored before. Radio Moscow reported regularly on its progress, such as its distance from Earth, speed and its position in the sky (in the constellations of Cetus and Pisces). The Venus probe was a worldwide event, the top story on the news pages in the press the following day and made the hoped-for global impact. The USSR had sent the first space probe to another planet, yet another 'first'.

Two communications sessions were held very soon after launch: one at 126,300 km on 12th February and another on 13th February at 488,900 km. A third took place on 17th February at a distance of 1.889m km while travelling at 3,923 m/sec, out over the Indian Ocean. Temperature inside the station was 29°C and pressure 900 mm. Scientific instruments reported back. The magnetometer recorded a faint interplanetary magnetic field of 3–4 nT (Earth's is between 24,000 and 66,000 nT), so this was close to zero. The solar wind detected by the First Cosmic Ship to the moon in January 1959 was confirmed.

From calculations made at this stage, ground trackers knew that the Automatic Interplanetary Station would not hit Venus – the injection out of parking orbit was simply not accurate enough. Trackers calculated a miss distance of 180,000 km. In addition to Yevpatoria, tracking stations were also being used in Irkutsk, Yakutsky and Mirny. Going through the telemetry stream and observations further, the miss distance was later refined to 100,000 km. Over a flight distance of 270m km, this was not bad, considering the early stage of development. Had there been time to fire the mid-course engine, then a hit would have been achieved.

'SABOTAGE IS NOT EXCLUDED'

The planned communications session with the probe on 27th February failed and contact with the probe was never regained. Radio Moscow announced on 2nd March that the interplanetary system had developed serious trouble in its telemetry system. During the interrogation on 27th February 1961, the high-frequency transmitter did not respond properly. In fact, fadeouts occurred. The next communication session was scheduled for 4th March, but a special test was brought forward to 2nd March, 'which revealed a malfunction in the permanent beacon transmitter.' Radio contact was therefore lost. An intensive investigation was now under way and Radio Moscow added the sinister note that 'sabotage during assembly is not excluded.' The station was due to reach 6.68m km the following day, 3rd March [7].

The large 76 m British radio dish at Jodrell Bank, near Manchester, had tracked Russia's moon probes in 1959, the Russians providing trajectory and signals data. In return, Jodrell Bank authenticated these Russian achievements, acted as a backup tracking station and sent tapes of the data back to Moscow afterwards. Now, the Russians returned to ask Jodrell Bank for assistance in their time of trouble. Jodrell Bank listened in for three hours on 4th March and seven hours on the 5th.

Yevpatoria commanded Venera 1 on the 17th May as it approached flyby, listening in along with Jodrell Bank. The British station did pick up signals on the 922.8 MHz frequency that day, briefly raising hopes that the probe might still be working, but they were spurious and not from Venera 1. Later, a team of scientists led by Dr Alla Masevich and Dr Jouli Khodareo travelled to Jodrell Bank from 9th to16th June and final attempts were made to contact the probe as late as 20th June. By way of a thank you, Mstislav Keldysh invited Bernard Lovell to visit the Soviet Union in June 1963. By coincidence, he arrived in Moscow in time for the rapturous celebrations that greeted the return from space of the first spacewoman, Valentina Tereshkova. A week later, Keldysh gave him a tour of the deep space tracking centre in Yevpatoria, the first Western visitor there [8].

Automatic Interplanetary Station: communications sessions

12 Feb	126,300 km and 488,900 km
17 Feb	1.889m km
22 Feb	Scheduled at 3.2m km; commands acknowledged
27 Feb	Scheduled, but 'fadeouts' reported
4 Mar	Scheduled at 7.5m km, failed

The exact point at which communications were lost is not certain and it is difficult to reconcile the contradictions between the various information available. The last *known* contact was 17th February, the third communication session (unlike most later missions, the Russians never gave a final total for this flight). Jodrell Bank picked up no signals after this date. The 19th is sometimes quoted as the last session, but that date is explicable insofar that the session on the 17th was not reported till the 19th. On the 19th, it was reported that ground stations were 'preparing' to contact the probe, indicating a session planned for the 21st or 22nd. The official history of OKB-1 states that the station responded to commands on the 22nd February, but does not say if it transmitted scientific or operational data then – presumably it did not.

It is significant that Burchitt and Purdy, travelling in Russia in summer 1961, give signal loss on 22nd February at the 3.2m km mark [9]. Granted that communications were intended to be a maximum of five days apart, it is certain that a session was intended between the 17th and 27th and the 22nd fits nicely. No successful session was ever reported on the 22nd and it seems that data were not recovered that day. The Russians probably did not panic and were convinced that the next timed session, five days later, would prove successful. It is difficult to make sense of the statement that 'fadeouts occurred' on the 27th and whether this meant that there was some form of signal recognition, albeit a faulty one.

The results of the investigation do not appear to have been published and we do not know if a saboteur was found. Granted what we know of the many technical difficulties that beset the early interplanetary programme, a more mundane explanation than sabotage may be plausible. Not for many years was a full explanation available. The initial explanations revolved around a failure in the thermal control system [10]. Later, it was determined that the unpressurized solar orientation system had given trouble from the start. It was decided to keep the station pointed at the Sun in a system of passive orientation, with the timer commanded to turn the spaceship around for communication sessions every five days and then re-orientate the station again. The receivers would be turned off between the sessions. The failure of communications was attributed to a timer failure, but mission controllers kept hoping that the timer would, somehow, still work. As a result of the loss of the mission, it was decided that in the future, timer design would be improved, the solar orientation system would be pressurized (and less likely to fail) and the receivers would never be switched off. The mechanical shutters had not worked well and would be replaced [11].

Whatever the investigation found out, the Academy of Sciences decided during the summer to move on to a new design. Some scientific information was obtained. For example, the station confirmed that solar wind, first detected by the First Cosmic Ship in 1959, extended far into space. The Automatic Interplanetary Station was a brave attempt and has a defined place in history as the first probe from our planet ever to travel to a distant world.

RE-DESIGN

The Automatic Interplanetary Station marked the end of the 1M and 1V (and 1VA) series. Korolev probably had little time to devote to Venus after February 1961, for the next month marked the final preparations for the flight into space of Yuri Gagarin. After the flight, he holidayed with his cosmonauts at Sochi, on the Black Sea, in early May, probably his first holiday. He was able to turn his mind to other issues in June. Even when the *Tizhuly Sputnik* was counting down, Korolev told his colleagues that the 1MV series would soon be over and that he wanted a new design, the 2MV. Preliminary sketches were done by Korolev at Baikonour in January 1961.

Korolev approved the 2MV series on 30th July 1961. A standardized bus was now developed, with four variants:

- 2MV-1 for Venus landers;
- 2MV-2 for Venus flyby;
- 2MV-3 for Mars lander;
- 2MV-4 for Mars flyby.

Soviet literature emphasizes that, with the 2MV series, the landers were 'detachable', implying that there was no system for separating the lander part of the first Venus probes. On the recommendation of the Institute of Microbiology of the Academy of Sciences, both were sterilized, a precaution against life forms from Earth being brought to Venus or Mars. Apparently, such a precaution had been overlooked in the haste of the 1960–1 launches. The landers were almost identical, the Venus ones having stronger shells and smaller parachutes, the Mars probes larger parachutes.

Following the problem with Venera 1, a new type of thermal regulation system was devised. New ground tests were done of the crucial manoeuvre of how to separate the lander from the mother ship. Systems were introduced to cool the landers: an airconditioner for the Martian lander and an ammonia-based system for the Venus

Vsevolod Avduyevsky

lander. Unlike the case with Venera 1, proper heatshields were designed so as to enable the descent craft to withstand the temperatures and strains of descent. These were designed by Vsevolod Avduyevsky (1920–2003), an expert in high-speed aero-dynamics and heat transfer from the Central Institute for Aviation, the Keldysh Research Centre and the Centre for Non-Linear and Wave Mechanics. One of the first things that they learned was that the best heatshields were not pointed but blunt, the latter forming a protective shock wave. Heatshields could be either ablative (the material would burn away, but slowly) or absorbing (they would soak up the heat, but not burn) or a combination. Epoxy resin was found to be the best component.

Altogether, six 2MVs were launched, three each to Venus and Mars in 1962. These were soon upgraded as the 3MV series by decree of the Council of Ministers and government of 21st March 1963 (#370-128). This decree approved the construc-tion of six 3MVs and three technology testers called Zonds, with the following designators:

- 3MV1 for Venus impact;
- 3MV2 for Venus flyby;
- 3MV3 for Mars impact;
- 3MV4 for Mars flyby; and
- 3MV Zond to test the technologies for all of them.

The idea of having a subset of technology demonstrators was Korolev's. His theory was that equipment for the interplanetary missions could be verified by missions launched during the long gaps in between the planetary windows, with additional tests during the real windows. These 3MV probes cover the interplanetary spacecraft launched from November 1963 onward, starting with the first Zond test of that month. The 3MV series was used for Mars missions in 1964 and Venus missions for 1964–72.

The decision of the Soviet government to compete with the Americans in the race for the moon (3rd August 1964, party and government resolution #655-268) had a small section approving a continuation of the interplanetary programme and, within that, a further six MV launches through to 1966, with a budget of 36,000 rubles (about $108m at then rates of exchange) [12].

2MV, 3MV series: key dates
30 Jul 1961	Approval of 2MV series by OKB-1
21 Mar 1963	Approval of 3MV series by government and party
3 Aug 1964	Continuation of 3MV series to 1966 by government and party resolution

Returning to the 2MV series, a new designer was appointed under Korolev's direction, G.S. Susser. Improvements included:

- New radio transmitter at 1 m wavelength using an omni-directional antenna.
- Duplicate transmitter for the main interplanetary part of the journey.
- On the flyby model, an impulse transmitter in the 5 cm wavelength.
- Solar–Earth sensors to point the high-gain antenna at Earth instead of the radio bearing.
- Area of solar panels increased to 2.6 m^2.
- Silver zinc battery replaced by cadmium nickel, capacity 42 amp/hr, providing for electrical reserve during planetary encounter.

These changes concentrated on the communications systems, indicating that this was where a problem lay with the first Venus probe. The 1.7 m high-gain antenna would transmit information at planetary arrival on the 5 cm, 8 cm and 32 cm bands. Small antennae on the solar panels would ensure transmissions at 1.6 m while still close to Earth, while the parabolic antenna would come into use as Earth receded into the distance. A system of pipes was introduced: instead of shutters which opened and closed and which could go wrong, hemispheric domes were installed at the end of each solar panel. These had coolant pipes, circulating ditolyl methane in the hot pipes and iso octane in the cool pipes to take away the excess heat or cold respectively. The domes were painted black and white for the appropriate hot and cold parts. The radio electronics, intended to ensure the high-rate transmission of data and photographs, were developed by Yevgeni Boguslavsky (1917–1969), deputy director of Mikhail Ryazansky's Institute of Space Device Engineering. Mikhail Ryazansky was an important person for the early interplanetary programme. Born 23rd March 1909 (os),[1] he was the engineer responsible for analyzing the control and guidance systems of the German V-2. In 1946 Korolev made him one of his council of chief designers and in 1955 he was put in charge of the NII-885 institute.

[1] Old Style, the calendar used before the Bolshevik Revolution, which ran twelve days behind the rest of Europe.

Yevgeni Boguslavsky

The 2MV flyby series carried a new type of camera, updated from that originally intended for the 1M Mars probes of 1960. These were the largest, biggest, heaviest cameras ever carried on a Soviet interplanetary spacecraft, 32 kg in weight, built by Ryazansky and the Leningrad NII-380 Scientific Research Institute of Television of Petr Bratslavets. The camera used two lenses (35 mm, 750 mm), took two types of images (square and rectangular), had two additional filters (infrared and ultraviolet), used 70-mm film and had the capacity to take 112 images, scanning them at either 68, 720 or 1,440 lines at a time, depending on the speed required. A high-power, high-data-rate impulse transmitter was again used, but it would still take up to six hours to send a full density picture. Mid course engines were installed, capable of making either one or two course changes.

Consideration was given to instruments to detect life on both Mars and Venus. Proposals were made by A.A. Imshenetsky of the Soviet Institute of Microbiology for taking soil into the cabin, adding in nutrients and then using instruments to measure the ensuing chemical reactions. The proposal got no further – it was probably too complex for these types of landers – though the Americans later used such a system on Mars in 1976 with Viking 1 and 2, with ambiguous results. One of the primary scientific aims of the 2MV series was to get temperature, pressure and density data back from the surface and for this sensors were developed by Vera Mikhnevich (b. 1919) of the Institute of Applied Geophysics in Moscow.

The 8K78 rocket was also improved, enabling the payload to rise from 640 kg to around 900 kg. Block A now had upgraded RD-108 engines, the 8D727K, with 5% more thrust and the third-stage block I was lengthened by 2.3 m. It was still called the 8K78.

Vera Mikhnevich

For the 1962 launch window in August–September 1962, three Venus probes were built: first, two landers and then a photo flyby. For the Mars window the next month (October–November), two flybys and one lander were built, six spacecraft in all. The tracking fleet set out across the oceans once more.

THE 2MV SERIES IN 1962

The experience of the *Tyzhuli Sputnik* returned to haunt the programme on the 25th August 1962. Although the first probe reached orbit, the fourth stage, block L, failed and the probe decayed out of Earth orbit. The system for firing out of orbit required the brief ignition of four small solid-fuel rocket motors to settle the fuel in the bottom of the tank (the Americans call this small manoeuvre 'ullage') and to point block L in the correct direction. Three of the motors fired, but not the fourth, leaving the stage out of alignment. The BOZ commanded ignition anyway, but the stage began to summersault after only 3 sec. This had become so violent that, after 45 sec, the engine was starved of fuel and cut out altogether. The stage fell out of orbit three days later.

On the second launching, on 1st September, a fuel valve failed to open on the block L engine, making ignition impossible and the probe fell out of orbit after five days.

Instruments on Venus 2MV landers, 25th August and 1st September 1962, with their designers

Chemical gas analyzer	Kirill Florensky
Temperature, density and pressure sensors	Vera Mikhnevich
Gamma ray counter	Alexander Lebedinsky
Movement detector	Alexander Lebedinsky
Gamma ray detector to measure radiation from surface rocks	Alexander Lebedinsky and Vladimir Krasnopolsky
Meteorite detectors	Tatiana Nazarova

This series was the first to carry gamma ray detectors and ultraviolet spectrometers. They were developed by Vladimir Krasnopolsky (b. 1938), who had graduated from Moscow State University in 1961 and was already an expert in planetary atmospheres. The gamma ray counter was developed by Alexander Lebedinsky (1913–67) of the Institute of Nuclear Physics at Moscow University, one of the Soviet Union's most expert astrophysicists. The meteorite detectors were developed by Tatiana Nazarova of the Vernadsky Institute. Instead of a dedicated instrument, the solar panels were wired in such a way to record any tiny dust particules that hit their exposed surfaces.

On the final launch of the series, the flyby on 12th September, there was a third-stage explosion at 531 sec, when the liquid-oxygen valve failed to close. It continued to supply oxidizer into the engine chamber, causing it to blow up and the third stage to break into seven pieces, an event detected by the Americans. Remarkably – and they did not observe this – the Venus probe atop its block L still entered parking orbit and attempted to begin its journey to the planet on the following revolution. When it did, the engine pump failed an eighth of a second into the burn, the engine cut out and it too was stranded in low orbit.

Vladimir Krasnopolsky

Alexander Lebedinsky

By contrast, the Americans had a much better experience with their first set of Venus probes. Much smaller, the Mariner spacecraft weighed only 200 kg and took the shape of a hexagonal base with solar panels. Mariner 1 veered off course, because an equation was entered wrongly in the computer programme and was destroyed. But Mariner 2 made a flawless departure away to Venus on 27th August, two days after the first of what would be three Russian Venus failures. Mariner 2's course to Venus was not that accurate and would have flown past the planet at 372,000 km, so a mid-course correction was carried out to adjust its path to a pass of 34,636 km. Mariner 2 made the first flyby of Venus that December, its instruments indicating a hot planet (425°C), a surface pressure of 20 atmospheres, a carbon dioxide atmosphere and a lack of water vapour. Contact lasted into January of the new year, when Mariner was 87m km distant.

Now it was time for the three Russian Mars probes to go: two flybys and a lander in that order. The guidance systems were rebuilt to prevent a recurrence of the problems experienced with the Venus missions, the weight penalties leading to some planned instruments being taken off. The following were the arrival dates scheduled:

Mars arrival dates (1963)

17 Jun	First flyby
19 Jun	Second flyby
21 Jun	Lander

In the event, this would have been wonderful timing, for the week in question saw the flight of Russia's first woman cosmonaut, but such a date had not been set the previous October.

During the first launching, on 24th October, the 890 kg probe reached Earth parking orbit of 218 km by 405 km and the engine fired for trans-Mars injection.

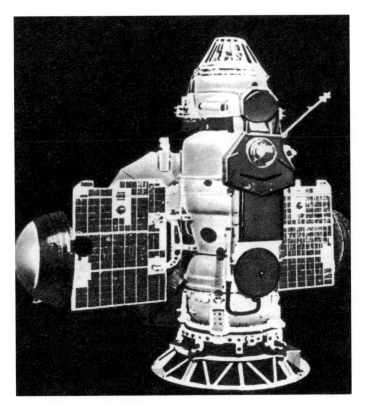

Coolant dome on solar panel (left)

Block L exploded after 16 sec when, after a hairline crack, the lubricant for the engine pump began leaking. The pump quickly overheated and jammed, causing the engine to blow up.

This was bad timing and location, for the Cuban missile crisis was at its peak and the explosion triggered off alarms in the American radar defence system, suggesting that the United States might be under Soviet missile attack. Thankfully, quick computer analysis showed that the débris pattern, comprising no fewer than 24 fragments, was quite different. The largest crashed into the atmosphere on 24th October.

The final probe, with the 305 kg lander, reached Earth orbit of 196 km by 590 km on 4th November. Vibration during the launch caused the fuse for the igniter of the third stage to pop out of its holder. As a result, the engine could not be fired. The Americans detected five pieces of débris, but it is not clear if this was due to a breakup or to the spacecraft and shrouds separating. It seems that the rocket crashed back into the atmosphere on 25th December 1962 and the lander on 19th January 1963. The lander had originally been overweight, a problem which chief designer Korolev solved in characteristic manner. Deciding which instrument should be taken off to save the necessary weight, he ordered the life detection experiment to be removed to see if it could detect life in the Kazakhstan desert. It couldn't, so it was taken off.

Instruments on Mars 2MV lander, 4th November 1962

Chemical gas analyzer	Kyrill Florenski
Temperature, density and pressure sensors	Vera Mikhnevich
Gamma ray counter	Alexander Lebedinsky
Movement detector	Alexander Lebedinsky

There were two basic problems underlying the high launch failure rate. First, the Russians experienced particular difficulty in mastering the fourth stage. This arose from the difficulty in igniting fuels in zero gravity. During the boost phase, fuels were normally all forced into the bottom of their tanks during the ascent. This was not the case for the fourth stage in parking orbit, where fuels had to be ignited when floating free in microgravity in parking orbit. Ways had to be found of forcing these free-floating propellants into their tanks for a normal, smooth ignition. Second, the 8K78 rocket was going through its early development phase and the rockets of the early space powers all had high early failure rates. More faults actually took place in the lower stages than in block L. Rocket scientists still had a lot to learn. A related problem was that the telemetry system on block L was, in retrospect, not sophisticated enough to determine the causes of failure with the certainty necessary to rule out a similar problem the next time. Block L did transmit data down to the tracking ships in the Gulf of Guinea, but in the form of a recorded data dump. It did not transmit systems data in real time, which would have permitted ground controllers to intervene effectively.

A further factor relevant to the high failure rate was the closed nature of the Soviet space programme. The Soviet Union was not the only country to experience high failure rates. An apt comparison is the early American programmes for the exploration of the moon, the Pioneer and Ranger series. Of the first nine lunar Pioneers (1958–60), only one even reached the moon, missing it (Pioneer 4), the rest either blowing up or falling back to Earth for insufficient velocity. Of the nine probes in the Ranger series (1960–5), the first six failed, sometimes for exasperatingly similar reasons as the early Soviet failures, such as timer faults. But, because the American moon programme was an open one, there was strong political pressure and incentives to rectify error and ensure the more economic use of scarce resources. Korolev and some of his designers were, arguably, shielded from some of these pressures.

MARS 1

The one bright spot in the five-out-of-six failure rate in 1962 was the flyby mission which left Earth orbit for Mars on 1st November 1962. Korolev was too ill to attend the launch and sent his colleague Boris Chertok to represent him.

Mars 1 was larger than its predecessor probe to Venus, weighing in at 894 kg. Mars 1 was a cylinder 3.3 m tall, with solar panels giving it a span of 4 m and a large umbrella-shaped high-gain antenna. Mars 1 carried hemispherical radiators on the tips of the solar panels with different external finishes so as to circulate colder or hotter water as required. The spacecraft had three radio transmitters, working on 1.6 m (omni-directional, like a television aerial on one of the solar panels), 32 cm and 5–8 cm wavelengths. There was a parabolic high-gain umbrella and a small semi-directional antenna located on the radiator. Their function was to transmit details of the condition of the spacecraft (temperature, pressure, battery seal) and then the readings on the scientific instruments, with the station commanded to report every two, five or fifteen days as required and depending on when Yevpatoria was in line of sight. A magnetometer swung out on a boom. The probe carried television cameras, meteoroid detectors and instruments for reporting on the atmosphere, radiation fields and surface of Mars.

Information was downlinked on 922.776 MHz and transmissions sent to the spacecraft on 768.96 MHz using the antenna attached to the solar heaters. Omni-directional backup antennae were located on the top end of the solar panels, using 115 MHz and 183.6 MHz. The 8 cm transmitter was intended to send pictures and data on 3,691 MHz when the spacecraft reached Mars. Designer of the radio system was Slava Slysh. A typical data transmission would interrogate the systems temperature, pressure and electrical current. Transmission were made for an hour or so at a time at two-day intervals for the first six weeks of the flight and thereafter every five days. Jodrell Bank followed the spacecraft for several months.

Mars 1 instruments and their designers

Television system for close approach to Mars	Petr Bratslavets
Spectro-reflexometer to detect organic compounds on the planet's surface	Alexander Lebedinsky
Spectrograph to measure the Martian atmosphere and ozone	Alexander Lebedinsky
Magnetometer to detect Martian magnetic field	Shmaia Dolginov
Gas discharge scintillation counter to detect radiation belts around Mars and cosmic radiation	Sergei Vernov
Cosmic radiation counter	Konstantin Gringauz
Radio telescope for cosmic waves in 150 to 1,500-m band	
Sensors for low-energy protons and electrons and ions	
Micrometeoroid detectors for cosmic dust	Tatiana Nazarova

The micrometeoroid counters were this time sensors set on the solar panels, sufficiently sensitive to detect any impacts on the panels. The camera system developed by Petr Bratslavets weighed 32 kg and was to take images through portholes using 35 mm wide-angle and 750 mm telescopic lenses, as well as take an ultraviolet image. The onboard scanning and transmission system could handle up to 112 pictures and transmit them at the rate of 90 pixels a second. Lebedinsky's infrared spectrometer, designed to analyze the vegetation on Mars and which had been left off on the 1960 probes, was on board this time. These instruments were sampled by a programmer and stored on tape. During a communications session, housekeeping data were transmitted first before the tape stored scientific data at high speed. As well as the downlink, commands could also be transmitted up to Mars 1 (uplink).

Mars 1

At first, it looked like another failure. Alarmingly, the first communications session reported that one of the two tanks of pressurized nitrogen, to be used for the mid-course correction engine and the orientation system, was leaking and within a few days was depleted. The outgassing not only spun the spacecraft out of control but deprived it of the fuel it needed for a mid-course manoeuvre. Another serious consequence of the leak was that it would be impossible to lock the high-gain antenna on Earth, making communications dependent on the low-powered omni-directional antenna. As a result of these problems, Soviet announcements heralding the first-ever mission to Mars were more muted than one might have expected. Radio Moscow stated that the initial path would bring Mars 1 to within 500,000 km of the planet. It had been intended to carry out a mid-course correction to bring the spacecraft to a pass distance of between 1,000 km and 10,000 km, the type of distance necessary for meaningful photography.

The telemetry system was able to indicate that falling nitrogen pressure in one tank was indeed the problem, giving the ground controllers at least some opportunity to rectify the situation and save something from the mission. After a week, ground control managed to use the remaining nitrogen in the other tank to spin the spacecraft around and achieve gyroscopic stabilization, pointing the station perpetually at the sun in such a way as to ensure that the solar batteries were constantly charged, so as to prolong the mission as much as possible. Now things began to improve.

The ground control station at Yevpatoria duly took signals from Mars 1, initially every two days and then five days. On 2nd November, the Crimean Astrophysical Observatory spotted the spacecraft, with its carrier rocket, as 14th magnitude stars and more than 350 photographs were subsequently taken following its position. As it travelled ever farther into deep space, Mars 1 detected and measured the solar wind and edges of the radiation zones around Earth. As it sped away from Earth, Mars 1 detected a third, outermost layer to the Earth's radiation belt at 80,000 km. Konstantin Gringauz's ion traps on Mars 1 now detected solar wind as it headed outward into the solar system. Shmaia Dolginov's magnetometer detected the same very weak interplanetary magnetic field as Venera 1. A solar storm was noticed.

Tatiana Nazarova's meteoroid detector was able to detect the dust clouds that Mars 1 flew through, clouds left behind by earlier comets. Mars 1 went through the Taurid meteor shower, recording a hit every two minutes between 6,000 km and 40,000 km distance and the same again when it met the stream between 20m and 40m km. Mars 1 continued to send back signals in the new year. By the end of January 1963, the craft was 27m km away from Earth and the time lag for acknowledgement of a signal was 4 min 47 sec. By 1st March 1963, the craft was 79m km distant.

On 2nd March, the Moscow newspaper *Pravda* reported that the strength of the signals from Mars 1 had declined, a warning of problems ahead. It appears that the orientation system broke down on 21st March and, as a result, the transmitters no longer pointed accurately toward the Earth. Communications were quickly lost, even though final telemetry indicated that all other systems on the probe were functioning properly. That date is now marked as the end of the mission. Mars 1 was 106m km away at the time, a new record for deep space communications and beating the then American record. It is possible that Mars 1 did take pictures of Mars when it flew past

Tatiana Nazarova

on 19th June 1963, but it transmitted them somewhere else in the galaxy. Radio Moscow revised the pass distance twice: first to 261,000 km and then to 193,000 km, which was probably the outcome – not bad, granted the lack of a course correction.

Mars 1 was still the first probe to Mars. Had communications lasted longer, a mid-course correction might have been carried out and this had the potential to close the miss distance to somewhere between 1,000 and 10,000 km. By the time of the loss of contact, there had been 37 communication sessions and more than 60 uplink commands had been transmitted. Mars 1 ended up in solar orbit of 148m by 250m km, circling the sun every 519 days. Useful scientific information had been returned from the in-flight instruments [13].

Science from Mars 1

Measurements of the distribution of charged particles in Earth's plasma envelope as the probe receded from Earth.

Measurement of the intensity of solar wind.

Detection of a solar storm of 600 million particles/cm^2/sec on 30th November 1962.

Intensity of radiation up 50% and 70% above level detected by the First Cosmic Ship in 1959, due to the higher stage of the solar cycle.

High level of collision with micrometeoroid dust near Earth, decreasing rapidly as craft proceeded out from Earth's orbit, increasing intermittently as the craft passed through the remains of meteorite showers.

3MV SERIES: 1964 VENUS MISSIONS

An investigative commission was set up by the Academy of Sciences during the first week of the mission when it looked as if the mission was lost. Even though Mars 1 was recovered, the investigators found many things that troubled them. The defective nitrogen valve was attributed to contamination endemic in the series of valves supplied. Other problems were so serious as to merit a complete redesign, so Mars 1 marked the end of the 2MV series.

The Russians may have banked on an increased launch rate leading to increased success in getting probes away to the evening star and the red planet. For the 1964 and 1965 windows to Venus and Mars, no fewer than eleven spacecraft were prepared for launch.

The 3MV series was broadly speaking similar to 2MV. The 3MV series was 3.6 m tall, 1.1 m in diameter, weighed up to one tonne and had two sealed compartments. One was a standard in-flight compartment, while the other was either a landing cabin (*pasadka*) or instruments for a flyby, such as cameras. Like Mars 1, the spacecraft carried a mid-course correction engine, the KDU-414, built by the Isayev design bureau, using for fuel unsymmetrical dimethyl methyl hydrazine (UDMH) and nitric acid. Thrust was 200 kg, specific impulse 272 sec giving a velocity change of 80 m/sec to 115 m/sec from a fuel load of 35 kg [14]. 3MV carried solar panels with a span of 4 m to charge the 14 volt, 112 amp/hr power system. The series had 32 cm transmitters and 1 m band receivers, with a 2 m tall diameter high-gain antenna for communications with Earth over long distances. Long-distance navigation was by way of an astro-orientation system which included not only the Earth and the Sun, but a reference star, normally Canopus. A shield was added to protect the sensors from extraneous light. Two tape recorders were carried for the retransmission of information.

Another important change in the 3MV series was in the camera system to image the planet during flyby. It was a radical improvement over the Mars 1 camera system. The new design was built by Arnold Selivanov of the Institute of Space Device Engineering. His system was comparatively minuscule compared with the cameras installed on the 1962 missions, coming in at a fifth the size, at 6.5 kg. The film used was 25.4 mm, able to hold 40 images and could be scanned at either 550 or 1,100 lines. Transmission rates were much faster and it would be possible to relay a high-density picture in 34 min rather than six hours. Additional infrared and ultraviolet filters were installed. They would take images through the three portholes on the side of the craft, sending them back through both a 5 cm and 8 cm band transmitter and recording the data on a tape recorder for retransmission.

The 3MV series, as noted at the start, included not only the 'Mars' designator, but also 'Zond'. Zond became one of the most problematic designations in the Soviet space programme. Probably because of the high failure rate in the 1MV and 2MV series, Korolev decided that a subset of the 3MV series should be used to develop the new technologies associated with the series. The Russians had a long history of using the 'Cosmos' label to cover for failures of spacecraft. Any mission that went wrong, especially lunar spacecraft stranded in low-Earth orbit, were called Cosmos (e.g., Cosmos 60). Since Cosmos was a programme for the exploration of space from Earth orbit, deep space missions could not plausibly be called 'Cosmos'. So when 3MV series were launched as 'Zond', it was presumed that the Russians were up to their old tricks and covering for failure [15]. In reality, Zonds were technology demonstrators and they were telling the truth. Later on, a version of the Soyuz manned spacecraft called Zond was used to carry out tests for a manned flight around the moon (Zond 4–8, 1968–70). Even recently, some veterans of the programme have insisted that, had these probes started better, they would have received Venera or Mars designators as appropriate, implying that Zond was originally an in-house designation only [16].

The first Zond was to fly past the moon, test out the camera system for the Mars flypasts planned the following year, enter solar orbit and maintain communications out to the distance of Mars' orbit. Apart from the camera system, this Zond carried cosmic ray sensors, a magnetometer, a solar X-ray experiment built by S.L. Mandelstam, a micrometeoroid detector and a solar plasma detector. Zond carried a star sensor, Earth sensor and sun sensor. For communications with Earth, there was a low-gain antenna and large parabolic antenna. Attached to the magnetometer was an antenna for communications with the landing module during its descent. Semi-spherical radiators were attached to the solar panels. The first Zond carried ion engines: the idea of long-firing, low-thrust electrical engines was first outlined by Tsiolkovsky in

1901 and the first such engine had been built on the ground in 1929. The first operational set of six thrusters was now built by Alexander Andrianov at the Kurchatov Institute.

The first Zond was launched on 11th November 1963 and reached parking orbit. But, 1,330 sec into the mission, block L lost its orientation and when ignition took place the spacecraft was fired in the wrong direction. It was named Cosmos 21 (it would have been called Zond 1 had it worked). This was the first time the USSR used the Cosmos designation for a failed lunar or interplanetary probe. The Cosmos series dated to March 1962 as a series of small scientific satellites built by the Yuzhnoye design bureau. Later that year, it began to be used as a cover name for spy satellites and it later became a cover name for any kind of mission the USSR wished to disguise. Had they thought to do so sooner, some of the failed Venus, Mars and lunar missions in 1962–3 might well have been called 'Cosmos' missions. Decades later, the practice was reversed: the scientific missions within the Cosmos series were civilianized and received their own, proper scientific designations. Nowadays, when the Russians launch a Cosmos, it is only a military mission.

For the Venus window of spring 1964, three spacecraft were prepared: two Zonds and a Venera. The second Zond was launched on 19th February 1964 and was designed to fly past the planet Venus. This was the second spacecraft to carry electric ion engines. But, following ignition on the third stage, a frozen pipeline burst and the stage exploded.

The second launch, a Venera designed to land on the planet, was set for 1st March, but was delayed four weeks due to failure to achieve a satisfactory integration of the spacecraft with the launcher. The lander was equipped with two 32 cm band transmitters, just in case one broke down during the descent. The Venera was eventually launched on 27th March, but block L lost its orientation in Earth orbit, never fired and was designated Cosmos 27. This time the telemetry's diagnostic systems had much improved and were able to explain not only this failure but many of the previous ones as well. Apparently, the gyroscope had been pre-programmed to inhibit an engine burn if block L deviated by more than a very limited degree from a pre-set orientation in the 70 sec before ignition. This orientation had been incorrectly set within far too narrow a range, preventing ignition when the rocket was actually within acceptable parameters. The frustrating thing was that the fault was so simple that the fix took only 15 min.

The third, this time a Zond, did get away on 2nd April and was publicly named Zond 1. The Russians published frequency details (922.76 MHz) and Jodrell Bank soon located the spacecraft. The Russians described it as a 'deep space engineering test', testing systems for interplanetary flight, giving its position in the sky but without mentioning the word 'Venus'. They did themselves a disservice, for the failure to acknowledge that Venus was the target created an air of mystery and deception. Indeed, on cosmonautics day, ten days later, Sergei Vernov stressed the important role of Zond 1 in studying high-level radiation belts above the Earth [17]. Very little information was released about Zond 1 at the time – the normally upbeat launch announcement was low key – and an illustration was not published until 1996, showing it to carry a landing capsule.

Zond 1 instruments (main spacecraft)
Radiation detector
Cosmic ray detector   Sergei Vernov
Magnetometer   Shmaia Dolginov
Ion detector   Konstantin Gringauz
Atomic hydrogen detector   Vladimir Kurt
Charged particle detector   Yevgeni Chukhov
Micrometeroid detector   Tatiana Nazarova

Zond 1 lander instruments
Temperature, density and pressure sensors   Vera Mikhnevich
Barometer
Thermometer
Radiation detector
Micro-organism detector
Atmospheric composition detector   Kyril Florensky
Acidity measurement detector
Electrical activity detector
Luminosity detector
Photometer   Alexander Lebedinsky and
                                                Vladimir Krasnopolsky

The charged particle detector was built by Dr Yevgeni Chukhov of the Theoretical
and Applied Physics Division of the Skobeltsyn Institute of Nuclear Physics of the
Moscow State University and was designed to measure proton fluxes above 30 MeV.
Measurements were made daily.

Vladimir Kurt

Zond 1 was the first Russian interplanetary spacecraft to carry a Lyman-α photometer to detect cosmic rays. Designed by Vladimir Kurt, they had already been tested on sounding rockets. On the lander was a a counter to measure gamma rays from the surface of the planet. Kirill Florensky's gas analyzer would measure the chemical composition of the atmosphere once it reached the surface. This was the first time a photometer had been carried on a lander. Because the cabin was set to land on the nighttime side of Venus, this was made sensitive to the lowest possible level of light (0.001 lux) right up to 10,000 lux.

Zond 1 used its mid-course manoeuvring engine for the first time on 3rd April when 563,780 km out from Earth. The need for the very early course correction suggests that the original blast out of Earth orbit was quite inaccurate and may explain the reluctance of the Russians to acknowledge it as a Venus probe. This was the first time the Russians had conducted a mid-course manoeuvre (although one had been intended for Mars 1 and Venera 1). The engine, called the KDU-414, was made by the Isayev Bureau. To get the engine facing the right direction, the spacecraft was orientated by little jet nitrogen thrusters. Nitrogen was kept in bottles at 320 atmospheres and fed into the thrusters at between two and six atmospheres. Once the gyroscope sensed that the spacecraft had achieved the appropriate velocity, it automatically cut off the thrust. Normal orientation used up about 300 g nitrogen a month.

It was soon apparent that Zond 1 was in even more trouble. One of the welding seams apparently cracked around the sensors, leading to a slow loss of pressure. By the end of a week, pressure inside the spacecraft was down to 1 mb – almost a vacuum. To complicate things, electrical systems short-circuited when exposed to the near vacuum. When the ion engines were turned on, they fired unevenly and had to be turned off. In the light of this, it is not difficult to understand why the Russians were reluctant to talk about the Zond.

Ground control struggled with the problem, devised a system for communicating with the probe through the lander and managed to stabilize the spacecraft. They still managed to carry out a second course correction on 14th May at 14m km distance out from Earth. This manoeuvre, of 50 m/sec, put Zond 1 on course to pass Venus at a distance of 99,780 km on 19th July, though this was still far too wide for it to deploy its landing capsule. The last announcement about the probe was on 19th May and communications were finally lost on 24th May. After the failure, it was decided to subject all space probes to X-ray tests to check that their welding seams were sealed tight. The British telescope in Jodrell Bank searched for signals as the spacecraft passed Venus that July, but with no result.

There seems to have been a good level of scientific return from Zond 1's scientific instruments. Much of the data on the conditions of interplanetary flight appears to have been lost, although records of protons detected appear to have survived.

IMPROVING THE LAUNCHER: THE 8K78M

Spring 1964 marked the point at which the Soviet Union began to phase out the 8K78 in favour of a 'modified' version, the 8K78M (hence the 'M'). This was first flown on the Zond on 19th February 1964. The new version was designed in a new place – no longer in OKB-1 in Moscow – but in its branch #3 in Samara, then called Kyubyshev and under the direction of designer Dmitri Kozlov. The plant was later called OKB-3 and is now called TsSKB Progress. The R-7 rockets had always been built there, but now derivatives were to be designed there as well. The 8K78M had a number of improvements:

- Replacement of the OKB-1, S1.4000, block L fourth-stage engine by an improved 11D33 built by the OKB-301 Lavochkin design bureau.
- Replacement of the block I third-stage engine by a new engine: the 8D715P and then the 8D715K built by Kosberg's OKB-154 and, from 1968, the 11D55.
- Improvements to the BOZ unit.

Later, over 1965–7, there were still some further modifications:

- The BOZ was lightened with the replacement of the four ullage motors by two.
- The nose fairing was redesigned, being lengthened by 67.5 cm.

For the 8K78 introduced in 1962 and flown over 1962–5, there were 20 launches, of which eleven failed. Even after the introduction of the 8K78M, the 8K78 continued to be used and was the launcher for Zond 3 in 1965, the two launchers running in parallel. The 8K78M was to continue the 8K78's bad launch record, with half of its ten launchings in 1964 failing. After that, the 'M' improved radically: only one failed in 1965 and in December 1965 the old 8K78 was retired. From 1964, the launcher was used to put communications satellites – called *Molniya* – into orbit and the launcher eventually acquired the title itself of the *Molniya* rocket.

3MV: 1964 MARS MISSIONS

The first Zond (November 1963) had been designed to test the technologies for the 1964 Mars window, which opened that November. Had either of the next Venus Zonds (February 1964 or Zond 1) succeeded, they would have added to the chances of success.

Three Zond spacecraft were available for the November 1964 Mars window, all flyby probes. Zond 2 was launched into a 153 × 219 km, 64.6°, 88.2 min parking orbit on 30th November 1964. The blast toward Mars was successful, putting Zond 2 into a wide solar orbit up to 1.52 AU out from the sun, taking 508 days. Zond 2 was the first interplanetary spacecraft announced to carry electric propulsion engines.

Disappointingly, the first communication session found that one of the two solar panels had failed and, as a result, Zond 2 was operating on only 50% of its normal electrical power. This was admitted at the time. Perhaps learning from the reaction to Zond 1, this time the Russians admitted an interplanetary objective and that scientific experiments would be carried out 'in the vicinity of Mars'. The failure of the solar panel to deploy was later traced to a cord breaking. When block L completed its burn,

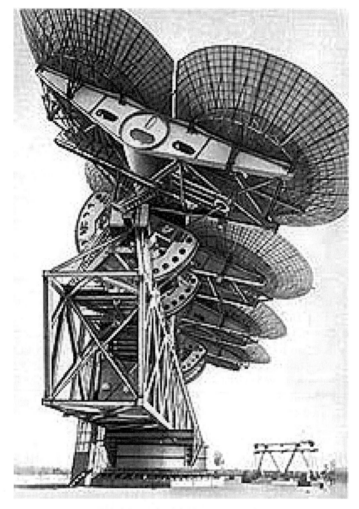

Zond 2 tracked from Yevpatoria

the shrouds were jettisoned, with cords on the shrouds supposed to pull out the solar panels as the Zond was ejected forward – but one of the two cords had broken.

It is possible that the Russians then decided to minimize in-flight operations and communications in order to save power for arrival at Mars. Next month, they used the spacecraft's plasma engines to shake the solar panel open, a manoeuvre which eventually succeeded on 15th December. By this time, though, it was too late to carry out the first mid-course manoeuvre.

Because analysis of the cause of this failure took some time, the Russians chose not to launch the two remaining Zond probes, with the risk – if not certainty – of maintaining their already appallingly high failure rate. These Zonds were put into storage while the problem was sorted out.

Zond 2 instruments
Radiation detector
Charged particle detector
Magnetometer
Micrometeoroid detector
Radio telescope
Cosmic ray detector
Solar radiation detector (Konstantin Gringauz)
Cameras

Zond 2 remains a problem mission for historians, many contradictory accounts being given of the progress of the mission. Originally, the mission was written off as a failure from an early stage and even some recent accounts say communications were lost after only a month [18]. Other accounts suggest that Zond 2 carried out a successful mid-course correction on around 17th February 1965, putting the probe on an accurate course to Mars and a pass distance of 1,500 km and that the last communications were on 2nd May 1965 [19]. The distance at this stage was 150m km, 50% farther than the record of Mars 1. There may be a way of reconciling these contradictions. We know that communications from Zond 2 were sporadic. Although mission controllers probably wrote off Zond 2 from an early stage as unable to carry out its original mission, they probably made some effort to keep in contact with and control the probe. Keeping contact with Zond 2 was probably not a priority in early 1965, for the Russians then began a major assault on the moon with a view to achieving a soft-landing. Jodrell Bank continued to pick up signals into February and the loss of signal announcement was made on 5th May [20]. Contacts seem to have continued up to then, even if the mission had been officially written off long before.

Despite the bad start and the low profile given to the mission by the Russians, the 950 kg Zond 2 was actually an interesting mission. First, it was the first inter-planetary spacecraft to fire its ion plasma engines successfully, six being carried and tested out on the 8th and 18th December. The six experimental plasma engines took the place of the 8 cm and 1 m wavelength transmitters and were an alternative to the gas engines.

Second, the probe took a long, slow curving trajectory. Western analysts did not pay much attention to this for some time, although it was odd. Most spacecraft before and since aimed to used the minimal fuel necessary to leave Earth, combined with the fastest transit time should communications prove unreliable. Zond 2's unusual behaviour was analyzed by Lepage, who commented astutely that Zond 2 'seemed to break all the rules' [21]. Zond's approach speed to Mars was slow, only 3.77 km/sec relative to the planet. The only merit in the trajectory was that it gave the slowest possible entry speed into the Martian atmosphere. He speculated that Zond 2 was to drop a landing cabin with a large parachute. The calculations of the time were that the Martian atmosphere had a density of 80 mb and a large parachute would have been able to lower a 380 kg lander to the surface safely.

In the event, Zond 2 was a photographic flyby mission, but the trajectory chosen was almost certainly designed to pave the way for a later mission that would follow this form of landing profile. On 6th August 1965, Zond 2 passed Mars at a speed of 5.62 km/sec, some 1,500 km from the planet. This was impressively close, even if the radio had failed. Very little was said about Zond 2 in the Soviet press. Details were not published until 1996 [22]. In appearance, it was like Zond 1, with a number of changes. The sensors were modified and the package now comprised a precision solar and star tracker, a guided parabolic antenna, Earth tracker and continuous solar tracker. On the bottom of the instrument module were three ports, one for a planet tracker, one for TV and one for photography.

Early Soviet long-distance communications distances

Feb 1961	Venera 1	3.2m km
Mar 1963	Mars 1	106m km
(May 1964	Zond 1	14m km)
May 1965	Zond 2	150m km
Mar 1966	Zond 3	153.5m km

ZOND 2 POSTSCRIPT

For the November 1964 window, the Americans followed their pattern for the 1962 Venus window and sent two spacecraft to Mars. These were small hexagonal-shaped box spacecraft with solar panels, an instrument frame and transmitters, weighing 260 kg. Although Mariner 3 did go into solar orbit, the spaceccraft was so damaged as to be useless because of a faulty nose shroud. Mariner 4 was placed on a flyby trajectory of 8,690 km and arrived at Mars in July 1965, a month before Zond. Mariner 4 sent back 22 pictures covering 1% of its surface. Its pioneering pictures were of poor quality by today's standards, but they ended the debate on the canals with unexpected finality. Mars was a dead world, like the moon, but even colder, and Mariner found no fewer than 70 craters. Most important, the atmosphere was very thin, about 8 mb. If a Zond had carried a lander with a large parachute, it would not have worked. This had serious implications for the development of future Mars landers [23].

Zond 2 had several other footnotes. The spacecraft planned for later in the launch window were eventually used, one for the next Venus probe (Venera 2). The other was eventually sent up the following 18th July and followed the mission scheduled for the probe originally launched in November 1963 which failed. Called Zond 3, it flew past the moon at 9,219 km some 33 hours after launch on a fast outward trajectory. This was the most successful of the early Zonds, although the Russians released few details until a press conference was held in Moscow a month after the start of the mission, when they had something to show publicly. They gave a good idea of how they intended to photograph the planets during flyby missions.

Lunar photography took an hour and began following an uplink command from Yevpatoria. Zond 3 swivelled to orientate itself with respect to the sun, Earth and

Zond 3

moon. As it passed the moon, Zond 3's f106.4 mm camera blinked away for 68 min at 1/100th and 1/300th of a second on aperture f8, taking 25 to 28 pictures of the farside of the moon, mainly that 30% part not seen by the Automatic Interplanetary Station in October 1959. Aligned with the cameras were lenses to measure the surface in infrared and ultraviolet light. Pictures were taken at 2 min 15 sec intervals and fed directly into the scanning chamber for development. They were transmitted back to Earth at two rates. First, at a quick scan rate taking 2 min 15 sec each, then at 1,100 lines, twice that of the American Ranger pictures taken of the moon a year earlier, in which case the signal took 34 min each. The idea seems to have been that – at planetary encounter – all the pictures taken would be relayed at fast speed. Ground control would then select the best and at its leisure command the spacecraft to transmit – in fine detail – the most interesting.

The pictures were first transmitted nine days after flyby using the narrow-beam, high-gain parabolic antenna locked on Earth at a distance of 2.25m km. First, the images were sent at high speed (67 lines/sec) and then high-density versions at low speed (2 lines/sec), each now taking a full 34 min to send. The pictures were re-transmitted in mid-August, again in mid-September and a third time on 23rd October from a distance of 31.5m km. The picture quality was excellent, showing lunar mountain ranges, continents and hundreds of craters sweeping past. Although Zond 3's orbit brought it out to that of Mars, an interception does not seem to have been planned. Zond 3 appears to have served its purpose of testing the photographic and transmission systems. It corrected its path on 16th September – when 12.5m km distant – with a velocity change of 50 m/sec. Plasma engines were again carried, but there is no information on how or when they were used. Course corrections were

Zond 3 over the moon

tested using the combined solar and stellar orientation system. The last reports of
Zond 3 were on 3rd March 1966 when it was 153m km distant from Earth, and contact
appears to have been lost at this stage, probably short of the distance of Mars orbit.
Zond 3 carried similar instruments to Zond 2, including the same cosmic ray detector.
Later that year, the Academy of Sciences published details of the detector's results
compared with those of Zond 1 and 2 and also the moon probes Luna 5 and 6.

Zond 3: scientific instruments
Two cameras.
Infrared and ultraviolet spectrometer.
Magnetometer.
Cosmic ray detector.
Solar particle detector.
Meteoroid detector.

END OF THE FIRST SERIES

Zond 2 marked the end of this stage of planetary exploration. The persistent difficulties in taming the 8K78 and 8K78M meant that many spacecraft were wasted. After
the 1MV series, only three had left Earth. Zond 1 had demonstrated the use of a mid-

course correction motor while Zond 2 had tested plasma engines. Much the most successful was Mars 1, which maintained communications to a distance of 106m km and obtained the first scientific return. Despite all their efforts, the Americans had been much more successful for a much more modest economy of effort and had obtained the first significant data from the two planets concerned. Granted the effort involved, this first stage must have ended in disappointment. An organizational change was called for.

REFERENCES

[1] Kotelnikov, V.A., Petrov, B.N. and Tikhonov, A.N.: Top man in the theory of cosmonautics. *Science in the USSR*, #1, 1981.

[2] Perminov, V.G.: *The difficult road to Mars – a brief history of Mars exploration in the Soviet Union*. Monographs in Aerospace History, no. 15. NASA, Washington DC, 1999.

[3] Oberg, Jim: *Red star in orbit*, 1980.

[4] Taubman, William: *Khrushchev – the man and his era*. Free Press, London, 2004.

[5] Varfolomeyev, Timothy: The Soviet Venus programme. *Spaceflight*, vol. 35, #2, February 1993.

[6] Salmon, Andy and Ball, Andrew: *The OKB-1 planetary missions*. Paper presented to the British Interplanetary Society, 2nd June 2001.

[7] Zygielbaum, Joseph L. (ed.): *Destination Venus – communiqués and papers from the Soviet press, 12th February to 3rd March 1961*. Astronautics Information, Translation 20. Jet Propulsion Laboratory, Pasadena, CA.

[8] Lovell, Bernard:
   – *The story of Jodrell Bank*. Oxford University Press, London, 1968;
   – *Out of the zenith – Jodrell Bank, 1957–70*. Oxford University Press, London, 1973.

[9] Burchitt, Wilfred and Purdy, Anthony: *Gagarin*. Panther, London, 1961

[10] Varfolomeyev, Timothy: The Soviet Venus programme. *Spaceflight*, vol. 35, #2, February 1993.
   Semeonov, Yuri: *RKK Energiya dedicated to Sergei P. Korolev 1946–96*. RKK Energiya, Moscow, 1996.

[11] Maksimov, Gleb Yuri: Construction and testing of the first Soviet automatic interplanetary stations, American Astronautical Society, *History* series, vol. 20, 1991.

[12] Siddiqi, Assif: A secret uncovered – the Soviet decision to land cosmonauts on the moon. *Spaceflight*, vol. 46, #5, May 2004.

[13] Ball, Andrew: *Automatic interplanetary stations*. Paper presented to the British Interplanetary Society, 7th June 2003.

[14] Clark, P.S.:
   – The Soviet Mars programme. *Journal of the British Interplanetary Society*, vol. 39, #1, January 1986;
   – The Soviet Venera programme. *Journal of the British Interplanetary Society*, vol, 38, #2, February 1985 (referred to as Clark, 1985–6).

[15] Gatland, Kenneth: *Robot explorers*. Blandford, London, 1974.

[16] Huntress, W.T., Moroz, V.I. and Shevalev, I.L.: Lunar and robotic exploration missions in the 20th century. *Space Science Review*, vol. 107, 2003.

[17] Russia plans new space flights, *Irish Times*, 13th April 1964.

[18] Huntress, W.T., Moroz, V.I. and Shevalev, I.L.: Lunar and robotic exploration missions in the 20th century. *Space Science Review*, vol. 107, 2003.

[19] Golovanov, Yaroslav (ed.): *Russians in space*. Compact Books, Moscow, 1997 (DVD); H.H. Kieffer, B.M. Jakovsky, C.W. Snyder and M.S. Matthews: *Mars*. University of Arizona Press, Tucson, 1992.

[20] Grahn, Sven: *Radio systems of Soviet Mars and Venus probes*. Posting at *http://www.sven-grahn.ppe.se*, 2005.

[21] Lepage, Andrew L.: The mystery of Zond 2. *Journal of the British Interplanetary Society*, vol. 46, #10, October 1993.

[22] Semeonov, Yuri: *RKK Energiya dedicated to Sergei P. Korolev 1946–96*. RKK Energiya, Moscow, 1996.

[23] Perminov, V.G.: *The difficult road to Mars – a brief history of Mars exploration in the Soviet Union*. Monographs in Aerospace History, no. 15. NASA, Washington DC, 1999.

4

OKB Lavochkin

*For forty years I have worked on jet propelled engines and thought that it would take
several centuries before we could take a trip to Mars. But time moves quickly and today
I am sure that many of you will witness the first flight beyond the atmsophere ... to the
moon and Mars.*

*– Konstantin Tsiolkovsky,
broadcasting from Kaluga on May Day 1933 to Red Square*

Zond 2 marked the end of the involvement of Korolev's design bureau in the Soviet
planetary programme. Korolev's design bureau, OKB-1, had almost single-handedly
begun and run the main programmes within the Soviet space endeavour. In August
1964, the Soviet Communist Party and government had resolved to compete head to
head with the Apollo programme to send a man to the moon. From there on,
Korolev's efforts were ever more focused on the man-on-the-moon programme,
adapting the N-1 rocket for lunar use and flying the Soyuz spacecraft. These were
huge undertakings. The decision was taken by Korolev and the government in spring
the following year to spin off several key aspects of OKB-1 to other design offices.
Communications satellites (e.g., *Molniya*) were made the responsibility of the NPO-
PM design bureau in Krasnoyarsk, for example. Here, all the interplanetary probes,
along with the troublesome block L, were given over to OKB Lavochkin.

OKB-301 LAVOCHKIN

The Lavochkin design bureau dated to July 1937 when it was set up by the aviation
designer Semyon A. Lavochkin (1900–60). During the war, it produced several
fighters for the Red Air Force, going on to build jets (the bureau built the

Semyon Lavochkin

swept-wing La-160 and the La-176 which broke the sound barrier, the first such Soviet aircraft to do so) and then air defence missiles, ramjets and drones. Semyon Lavochkin had wanted to move his bureau more into the space business and in 1959 had recruited a team of 15 young designers to look at moon probes. He made a point of recruiting graduates from the Moscow State Technical University, also known as the Bauman Institute and from the Moscow and Kazan Aviation Institutes.

Semyon Lavochkin died in June 1960 and that December part of the bureau was reassigned to Sergei Korolev. Several members of the young designer team were relocated to OKB-1, where they contributed to the design of the 1MV and 2MV series. The main part, making military and anti-ship missiles, went to Vladimir Chelomei where it became OKB-52 Branch #2. There was not much left.

Now on 2nd March 1965, Lavochkin's old bureau was reconstituted as GSMZ *imeni Semyon Lavochkin* (literally, state union machine building plant dedicated to the memory of Semyon Lavochkin), known colloquially as the Lavochkin design bureau. The term OKB-301 also continued in use. Appointed to lead the new OKB was general designer Georgi Babakin. Fifty-year-old Georgi Babakin was an unusual man, self-taught, with a healthy suspicion of formal education. Born in Moscow on 31st October 1914 (os), he developed an early passion for radio electronics, becoming senior radio technician with the Moscow Telephone Company in 1931. He was drafted into the Red Army's Proletarian Infantry Division in 1936 where he was a radio operator for six months before being dismissed for ill health. He returned to school, where he completed his exams, joining the old Lavochkin design bureau during its plane-making days in 1951 as an autopilot expert, rising to deputy chief designer. He eventually took a university degree in 1957 [1]. When Semyon Lavochkin died, Babakin went to work for Vladimir Chelomei, before being called to head up the recreated OKB-301.

In April 1965, Sergei Korolev made his first and only visit to the Lavochkin design bureau. He met all the senior design staff, formally handed over the OKB-1 blueprints

Georgi Babakin

to them, made clear the heavy duty now incumbent upon them and warned them that he would take the projects back if they did not perform [2]. Lavochkin's experience of producing military aircraft stood to its advantage, for the company put much emphasis into ground-testing and cleaning bugs out of the system beforehand. Indeed, Korolev admitted to the Lavochkin designers that much of the work in OKB-1 on interplanetary probes had been rushed and that they had suffered from greater attention being given to other projects. The official history of OKB-1 admits that the probes were probably too complex and ambitious, the long missions requiring a level of quality and reliability that was beyond the standards of Soviet precision engineering of the time.

Few people seem to have moved across from OKB-1 to Lavochkin. One who did was Oleg Ivanovsky. Another radio enthusiast, he was a cossack cavalryman during the war but was so badly wounded that at war's end he was registered permanently disabled, facing a grim future without work or, more importantly, worker ration cards. An old friend managed to get him work in OKB-1 where his radio skills were quickly appreciated. Korolev gave him a key role in the radio instrumentation for Sputnik, the 1959 moon probes and then Vostok, personally accompanying Yuri Gagarin to his cabin. When the new Lavochkin company was set up, Korolev found him a post as deputy chief designer, second only to Babakin [3]. Another important personality of the new bureau was Vladimir Perminov (born 1931), a graduate of Kazan Aviation Institute, a subsequent specialist in spaceplanes and now appointed to lead the redesign of the 3MV series.

A casualty of the change was Gleb Maksimov, the designer of the IMV, 2 MV and 3MV series, whose involvement in interplanetary research ended at this stage. He stayed with OKB-1 for another three years and then joined the Institute for Space Research, taking a doctorate there and going into teaching at the Moscow High Technical School.

Oleg Ivanovsky

THREE-PART REORGANIZATION: LAVOCHKIN, MINISTRY, INSTITUTE

The handover of interplanetary probes to the Lavochkin design bureau was part of a much larger reorganization of the Soviet space programme, one associated with the change of power in the Soviet Union in October 1964. Until the change, decision-making in the Soviet space programme had not followed a formally defined institutional pattern. Although the Soviet Union was portrayed in the West – and indeed liked to portray itself – as a command economy in which the party and government led and issued instructions for the good of society, economy and science, the reality was quite different. The early space programme had been led 'from below', with chief designer Sergei Korolev putting proposals up to the political leadership of party and government, then very much concentrated in the hands of Nikita Khrushchev. In the case of the early space programme, Sergei Korolev had presented memoranda to government and party, in the process spending considerable time winning over the leading figures in the Academy of Sciences, the Aviation Ministry and Defence Ministry. Once he had got all the key figures on his side, the government and party were normally ready to approve the necessary resolutions that authorized programmes to proceed. By contrast, other countries organized space efforts through national space agencies, such as NASA in the United States (1958) and CNES in France (1962). Western analysts looked hard for a similar agency in the Soviet Union, but failed to find one, presuming that it was just too well hidden from view. They could not imagine that the reason was that it did not exist.

For the early years Korolev had a free run and, in the apparent order and logic of the early Soviet space programme, owed much to his ability to win round the scientists, ministries, party and government to his vision and priorities. It was not long before rival designers found that this was a game in which they could play too. Some, like Vladimir Chelomei and Valentin Glushko, found themselves well suited to the strategies and intrigue involved. Far from being a clinical command-and-control system, the Soviet 'system' was actually a disorganized, even chaotic playing field,

Sergei Korolev and Mstislav Keldysh

in which warring factions contended, rivalries went untamed and decisions could be unmade and remade with alarming rapidity and considerable, wasteful inconsistency. As far back as 1959, both Korolev and Keldysh had argued that there should be a dedicated body for space research and Keldysh had repeated the proposal in 1963.

There were many reasons which prompted plotters to move against Nikita Khrushchev, who they overwhelmed in a bloodless *coup* – his concentration of personal power was one of them. The Soviet Union must, they felt, be protected against the kind of adventurism that had led, domestically, to several economic and agricultural disasters and in foreign policy to the Cuban enterprise. As a result, the personalities of government and party were separated (Leonid Brezhnev becoming party secretary, Alexei Kosygin prime minister), a style of 'collective leadership' was instituted (or reinstituted), power was diffused and ministries were reorganized. This process took several months and concluded in the spring and summer of 1965. Strange though this may seem through the lens of history now, the younger, fitter, brighter Leonid Brezhnev promised a more pragmatic, rational, technocratic, competent, consistent, collegial leadership, infused with fresh energy.

As part of this reorganization, it was decided to establish a ministry responsible for the space programmme, making it a defined area of governmental and ministerial responsibility and establishing a layer of decision-making between the design bureaux on the one hand and the government and party on the other. The new ministry was entitled the Ministry of General Machine Building. 'General Machine Building' was a sonorous title specifically chosen to hide its true significance. Such titles were an endemic feature of Soviet government (the Ministry of Medium Machine Building ran the nuclear industry; but the USSR confused the issue with some titles that were vaguely truthful, such as the Ministry of Heavy Machine Building which really was responsible for cranes and excavators).

The various design bureaux, which had been assigned to a range of different ministries (e.g., aviation, defence), now came under the Ministry of General Machine

Sergei Afanasayev

Building. Appointed to lead it was the Minister of General Machine Building, Sergei Afanasayev. A 47-year old defence economist, he was nicknamed 'the big hammer'. Although disliked and feared by some in the space industry, he was respected for his energy, his ability to get things done and to clear bureaucratic hurdles. The true nature of the Ministry of General Machine Building was never revealed and neither Sergei Afanasayev, nor the other chief designers, had any kind of public profile.

The new ministry came into existence on 2nd March 1965, on the same day as the reestablishment of the Lavochkin design bureau. The final, third part of the same reorganization took place later that summer. The Korolev and Keldysh proposals for a space research body were at last heeded and an Institute for Space Research (IKI, or *Institut Kosmicheski Izledovatl*) was brought into existence. The first director was aero-dynamicist Georgi Petrov; ideally, the institute would help to bring a more rational, orderly framework to the planning of the Soviet space programme, one in which scientific considerations were uppermost. The new government decided to locate IKI within the Academy of Sciences. Even in Stalin's time, the Academy of Sciences had retained its international links and was an acceptable public face for the space programme at home and abroad.

It took some time for the new institute to build its role but, by the early 1970s, IKI was able to define the research priorities of the Soviet space programme, not least the interplanetary programme, commissioning the design bureaux and overseeing the missions. The founding date for the institute was 14th July 1965. It was built in southwestern Moscow where an old village had formerly stood. The Academy of Sciences institutes for mathematics and chemistry were also moved there into two large office buildings, surrounded by flats for the workers there. Twenty years later, IKI's staff had grown to 1,500 people. Another aspect of the establishment of IKI was that it brought coherence to the instrumentation flown on Soviet interplanetary probes. Vasili Moroz, Vladimir Krasnopolsky and Leonid Ksanformaliti were put in charge of instrumentation for future Mars and Venus probes [4].

The changes instituted in March 1965 were to prove much less effective than the new, neat organigrams suggested. In some respects, the new ministry provided just another battleground on which warring factions could contend. In the moon programme especially, the quality and continuity of decision-making did not improve and was to cost the Soviet Union the race to the moon. The new system did not reach a level of stability until a fresh upheaval in May 1974, but, as we shall see, remained punctuated by discontinuities. In the interplanetary programme, it did permit the emergence of IKI which in the fullness of time facilitated a growing level of international collaboration.

LAVOCHKIN'S REDESIGN

The next window for a mission to Mars was 1967. This now became the responsibility of the Lavochkin design bureau. The bureau took the view that it did not wish to duplicate the experience of America's Mariner 4, but to make a substantial advance beyond. Informed by Tikhov's thinking and the wisdom of the day, the existing Mars lander in the 2MV and then 3MV designs had been based on the assumption that the Mars atmosphere was between 0.1 and 0.3 that of Earth and certainly not less than 80 mb. Over 1965 the bureau explored various options of how to adapt the 3MV lander to ascertain whether it could make a soft-landing in an atmosphere now shown by Mariner 4 to be much thinner. It concluded that, even with a big parachute, the descent would be so quick – only 25 sec – as to provide only a limited cushion of atmosphere as the craft descended, with the risk that it would land too hard to survive. In October 1965, the 3MV design was abandoned for Mars, though continued for the much denser atmospheric conditions of Venus. In the case of Venus, OKB Lavochkin made a number of immediate changes to the 3MV series, but Venera 4 was constructed using documentation supplied by OKB-1 [5]. Babakin wanted to make a break with the past by calling the probe the V-67 (after the year of launch), but Korolev's old designator system somehow hung on and the designator 4V-1 was also used (and subsequently 4V-1M and 4V-2). In the case of Mars and later Venus, a new generation of spacecraft would be required.

On 22nd March 1966, Georgi Babakin gave his comments on the proposals by his designers for a new generation of spacecraft to explore Mars and Venus. Like the 1MV, 2MV and 3MV designs, the new design would incorporate flybys and landers but with the additional possibility that the mother ship could enter orbit over the planet (parameters of 2,000 km to 40,000 km were suggested). The Mars lander must be specifically adapted to suit the low density of the planet's atmosphere. Georgi Babakin identified the issue of data transmission as a key concern, specifying a transmission rate from flyby or orbiter of at least 100 bits/sec and from the lander of 4,000 bits/sec. They would be significantly larger.

Ambitious scientific objectives were set for the new generation of spacecraft. In the case of Mars, these were listed as to:

- soft-land, taking images of the surface and vegetation;
- measure prevailing conditions on the surface (temperature, pressure, wind);
- measure the parameters of the soil – composition, temperature, density;
- detect micro-organisms;
- from Mars orbit, study the atmosphere, compile a radiothermal map, image the Martian moons and image the surface to determine the nature of 'seas', 'canals' and seasonal changes.

In the case of Venus, the objectives were listed as to:

- profile the atmosphere by altitude for temperature, pressure, chemical composition and illumination;
- image the surface;
- test the mechanical properties of the soil;
- study the atmosphere and compile a radiothermal map;
- image Venus from its orbiter;
- detect micro-organisms.

As the planetary objectives showed, neither Mariner 2 nor 4 had killed off a residual belief in the life forms imagined by Gavril Tikhov. In addition, the spacecraft were to study interplanetary space *en route* to their destinations for solar and space radiation, meteorites and the magnetic, electrical, gravitational and radiation environment.

The scientific profiling of the atmosphere was assigned to Kiril Florensky (1915–1982), the Soviet Union's foremost planetary geologist. He had joined the Vernadsky Institute for Geochemistry in 1935. He led, in 1958, an expedition to Siberia to uncover the mystery of the 1908 impact in Tunguska, now believed to be a small comet and he became a leading theorist on planetary surfaces and their atmospheres. While Florensky focused on the composition of the atmosphere, another scientist, Viktor Kerzhanovich (b. 1938) of Moscow State University, developed a system to measure its turbulence and rotation.

Kiril Florensky

Viktor Kerzhanovich

Development of this series coincided with the building by the Lavochkin bureau of a new generation of moon probes, Luna 15 to 24. They had a broadly similar size and weight and were designed with similar purposes in mind for orbiting and landing. The propulsion systems and fuel tanks shared a number of common features.

NEW SPACECRAFT: NEW ROCKET – THE UR-500 (8K82)

The 8K78 and 8K78M were able to send only about a tonne to Venus or Mars. Had miniaturization or micro-electronics been available, this would not have been a serious limitation, but at the time the one-tonne ceiling set significant limits to what could be achieved, especially in the case of Mars. A larger rocket was imperative for more ambitious planetary missions.

The new Soviet rocket dated to October 1961, when Russia detonated over the northern Arctic island of Novaya Zemlya its first 58-megatonne thermonuclear superbomb. This bomb was carried aloft and detonated by a Tu-95 bomber, but there was no way these ageing propellor planes could reach – never mind drop – their cargoes on New York. For this purpose, a new powerful rocket was required. Nikita Khrushchev turned to Vladimir Chelomei (the man who gave him his military rocket fleet) who now promised to built him an ever bigger rocket, called the Universal Rocket (*Universalnaya Raketa*) 500, UR-500, with the design code of 8K82. Khrushchev was soon bragging about the Soviet Union's new 'city-buster' rocket.

In the event, the UR-500 was never taken into armaments and was cancelled as a military project very early on by the government of Leonid Brezhnev when it took power in October 1964. The UR-500 survived as a space rocket, converted to civilian use. Vladimir Chelomei astutely persuaded the Kremlin that the UR-500, with suitable upper stages, could send a small manned spacecraft round the moon, and it was

Vladimir Chelomei at the blackboard

for this purpose that it was subsequently developed. Georgi Babakin saw the potential, with the addition of more stages, of developing the UR-500 for interplanetary spacecraft and that it could quadruple the interplanetary payload at a stroke. As was the case with the R-7, several variants of the UR-500 were later built, being used to dispatch space stations, communications satellites, deep space probes and navigation satellites.

Chelomei's rocket used nitrogen tetroxide and unsymmetrical dimethyl methyl hydrazine (UDMH). These were the nitric acid fuels with which Valentin Glushko had first experimented in the early 1930s. They had the advantage (originally important for military rockets) of being storable fuels that could be kept in their tanks for long periods of time before take-off, did not require cooling facilities, were powerful and built up thrust very quickly. On take-up, nitric acid fuels give off a tell-tale, distinctive, orange–brown flame.

The engine of the first stage was built by Valentin Glushko's OKB-456 and was designated the RD-253 engine. The RD-253 was the most advanced rocket engine in the world for 20 years or more. They recycled their exhaust gases to create a closed-circuit turbine system. Each engine weighed a modest 1,280 kg, but pressures of hundreds of atmospheres were obtained. The turbines went round at a fantastic 13,800 revolutions a minute. Temperatures reached $3,127°C$ in the engine chambers and their walls had to be plated with zirconium. Their thrust was 2,795 m/sec at sea level and 3,100 m/sec in vacuum. Equally significant was the clustering of fuel tanks around the side of the bottom stage. The 4.1-m diameter restriction of the rail system could make any powerful rocket far too slim to be viable. What Chelomei did was develop the main core as the oxidizer only, within the 4.1 m limit and later attach the fuel tanks to the side of the rocket. They were built separately, transported separately from Moscow to Baikonour and then attached in finishing hangars beside the pad in Baikonour. With the tanks attached, the diameter of the new rocket on the pad was 7.4 m.

Chelomei's rocket was and still is built in the Fili plant in Kaliningrad, now known as Korolev. This was an old automobile factory, taken over by the Bolsheviks to build German Junkers planes in the 1920s and then Tupolevs. It became Vladimir Chelomei's OKB-52 in 1960 and its descendant is now known as Khrunichev, an affiliate of Lockheed Martin. The new rocket required the building of fresh pads to the northwest of the cosmodrome. Two sets of double pads were built, called Area 81 and Area 200. Each had a left pad and a right pad (81L, 81P, 200L, 200P). The rocket was brought down to the pad on a train trailer and then erected into the vertical position. Around the pad were 100 m tall lightning conductors and four 45 m tall floodlight stands. A shallow flame trench took away the rushing roar of the engines firing at take-off from both sides.

It took Chelomei and his OKB-52 fewer than two years to design the UR-500 (1961–3) and fewer than two years to build it (1963–5), all the more remarkable granted its cancellation as a military weapon. The design was subject to rigorous ground-testing, and Chelomei refused to rush things. The UR-500's first mission on 18th July 1965 went like a dream, lofting the first of a series of four large cosmic ray satellites over 1965–8. They were called Proton and the first, Proton 1, was the largest scientific satellite ever launched up to that point, weighing in at no less than 12 tonnes. Of the UR-500's first four launchings, only one failed, making it the most promising rocket of its day – a deceptive début, as events were to transpire. The first launch was well publicized. Although the first UR-500 had the new name of *Herakles* written on its side, the Soviet authorities chose to call it the more mundane Proton rocket after the first satellite it put in orbit. Possibly because of its military origins, the Russians kept back details of Proton for well over 20 years until, as we shall see, the VEGA missions in 1984.

Despite its promising start and despite Chelomei's thoroughness, the Proton was, like Korolev's 8K78, to have an exasperating development history. Of its next 24 launches, no fewer than eleven failed. At the time, nobody would have credited that Proton would go on to become one of the most reliable rockets in the world and Proton launchings passed the 315 mark in 2005. Although there were occasional final-stage failures with block D, lower-stage failures became most unusual (there were two in the 1990s, when quality control in the manufacturing plant slipped during the period of greatest economic difficulty). A new, much more powerful version of Proton was even introduced, the Proton M, in 2001.

And it was perfect for the new planetary missions being planned by Georgi Babakin in the new OKB-301 *imeni Semyon Lavochkin*.

The Proton rocket

UR-500K (three-stage version, with block D fourth stage)
Length   44.34 m
Diameter   4.1 m

First stage (block A)
Length   21 m
Diameter   4.1 m
  with tanks   7.4 m
Engines   Six RD-253
Burn time   130 sec
Thrust   894 tonnes
Fuels   UDMH and N_2O_4
Design   OKB-456 (Glushko)

Second stage (block B)
Length   14.56 m
Diameter   4.1 m
Engines   Three RD-210, one RD-211
Burn time   300 sec
Thrust   245 tonnes
Fuels   UDMH and N_2O_4
Design   OKB-456 (Glushko)

Third stage (block V)
Length   6.52 m
Diameter   4.1 m
Engines   One RD-213, one RD-214
Burn time   250 sec
Thrust   64 tonnes
Fuels   UDMH and N_2O_4
Design   OKB-456 (Glushko)

Fourth stage (block D)
Length   2.1 m
Diameter   4.1 m
Engine   One 58M
Thrust   8.7 tonnes
Length   6.3 m
Diameter   3.7 m
Fuels   Liquid oxygen and kerosene
Design   OKB-1 (Melnikov)

RUSSIA REACHES THE EVENING STAR

Until the new designs could be ready, for they would take several years, Georgi Babakin decided to persevere with the 3MV series of probes originally developed by OKB-1.

Four probes were prepared for the autumn 1965 Venus window, including one Zond left over from the Mars window a year earlier. This time the first two – one a flyby, one a lander – left Earth orbit for Venus, the first time more than one launch per window had succeeded. Of the rest – one a lander, one a flyby – neither left Earth. The lander was stranded in Earth orbit by a fourth-stage failure on 23rd November 1965 when the fuel pipeline of block I tore at 528 sec, causing a combustion chamber explosion. Although block L and the Venus probe entered orbit, the chamber explosion had caused it to tumble: it was not able to fire out of orbit and was called Cosmos 96, eventually crashing to Earth on 9th December. On 26th November, the fourth launch had to be abandoned due to difficulty with the launcher and could not be fixed before the window expired. This window saw a change in the nature of the parking orbit used. Hitherto, the parking orbit revolved at 65° to the Earth, as did manned flights. From now on, the lower inclination of 51° was used, giving some extra payload capacity. This remains the standard inclination to this day – it is the orbit of the International Space Station, for example.

Venera 2 departed Earth parking orbit on 12th November 1965. Weight was 963 kg. The first to be named 'Venera' at launch ('automatic interplanetary station' was relegated to a subtitle), the planned course correction was cancelled when it was apparent that the probe would come within 40,000 km of the planet during its scheduled flyby on late February 1966. There were high hopes of television pictures during the daylight side flyby, using the system now proven by Zond 3. The camera system was the same, but a 200 mm lens was installed instead of the normal 106.4 mm lens. Twenty-six communications sessions had been carried out by the time of Venus approach. In reality, the blast out of parking orbit had proved to be exceptionally accurate and final flyby distance was only 23,950 km. Once it passed close by, Venera 2's cameras and instruments would swing into action. Alexander Lebedinsky had installed two spectrometers to scan the surface and, to make sure that they worked, they had already been tested out on Earth-orbiting Cosmos satellites.

Officially, the final communication session, the 27th, did not take place just at the moment of Venus flyby. Contact with Venera 2 had been lost just beforehand, it was frustratingly admitted, but there had been a close pass. In fact, things were a little more complicated. During the last communications session before encounter, as the probe approached the planet, Venera 2 reported a sharp increase in temperatures on board. Ground control sent the uplink command for the flyby imaging and experiments to commence, but the actual communications session was of poor quality and the command was not acknowledged. Once the flyby had taken place, ground control attempted to establish contact for the planned communication session and to download the scientific data and images collected during flyby. Contact with the probe was never regained and it was declared lost on 4th March. Post-mission investigation suggested that the radiators had failed as the probe approached the planet, causing failures in the command, receiving and decoding systems. The scientific instruments may well have done their job and taken the first-ever pictures of Venus, but they could not be transmitted.

As for Venera 3, which was a lander, it had left Earth four days later. In contrast, the insertion toward Venus was not as accurate and would mean a miss

Venera 2

of 60,550 km. A series of calculations, 13,000 in all, were undertaken to try to rectify the trajectory of the 960 kg craft. A course manoeuvre of 19.68 m/sec was duly carried out on 26th December when it was 12.9m km out from Earth to aim the probe dead centre on Venus and this manoeuvre duly increased its speed to do so.

Venera 2–3 in-flight instruments and their designers

Magnetometer	Shmaia Dolginov
Cosmic rays detector	Slava Slysh
Low-energy charged particle detector	Krupenio
Solar plasma detector	Konstantin Gringauz
Micometeorite detector (not Venera 3)	Tatiana Nazarova
Hydrogen and oxygen spectrometer	Vladimir Kurt
Radio detectors in 150 m, 1,500 m and 15 km bands	

Venera 2 flyby instruments

200 mm camera	Arnold Selivanov
Infrared spectrometer	Alexander Lebedinsky
Spectograph	Alexander Lebedinsky and Vladimir Krasnopolsky
Spectrometer	Alexander Lebedinsky and Vladimir Krasnopolsky

Venera 3 lander instruments

Temperature, density and pressure sensors	Vera Mikhnevich
Gas analyzer	Kiril Florensky
Photometer	Alexander Lebedinsky and Vladimir Krasnopolsky
Movement detector	Alexander Lebedinsky and Vladimir Krasnopolsky
Gamma ray counter	

The intention of Venera 3 was to detach the bucket-shaped 383 kg, 90 cm diameter descent capsule into the planet at an angle of between 43° and 65°, where it would descend under parachute. This was in the small container on the bottom. Venera 3 reached the planet on 1st March 1966, after a 105-day journey. When it did, it became

Venera 3

the first body from Earth to reach the surface of another world, depositing there pennants of the Soviet Union and Lenin.

The impression was given that Venera 3 had transmitted close to Venus arrival and that only a last-minute failure had robbed them of success with Venera 2. In reality, communications with Venera 3 had been lost much earlier, on 15th February, but this was buried deep in a *Pravda* article. The announcement of the landing was made on the assumption (probably correct, but in absolute terms unverifiable) that Venera 3 was aimed so accurately that it would hit Venus. Official announcements explained, in some detail, the mathematical basis of these assumptions and they appeared to be well grounded. The USSR said that it had made, in the course of 63 communication sessions, no fewer than 5,000 calculations of velocity and 7,000 of angular coordinates. Even allowing for a reasonable margin of error, the result would still be a hit, probably somewhere in the area 20°N to 30°S, 60°E to 80°E. We do not know if the descent craft was separated or not, though the automatic command system probably meant that it was. Reaching Venus was hailed as a great success, which it was, but the joyous outcome concealed the fact that Russia had now launched no fewer than 18 probes to the planets and none had yet fully completed its mission.

Post-mission analysis found that Venera 3 suffered similar problems of thermal control and it may be the case that Venera 2 was the hardier of the two probes. The painting system for the hemispheric domes had failed, one of a series of problems of poor manufacturing standards. Whatever the cause, it is certain that measures were taken to improve the thermal protection system thereafter. By this stage, Soviet scientists had finally come round to the idea that Venus was a hostile, hot planet. President of the Academy of Sciences Mstislav Keldysh wrote an article in *Pravda* on 6th March to mark the Venus landing, where he expressed his personal belief that surface temperatures were in the order of 300°C to 400°C and that the planet had succumbed at some earlier stage in its history to the greenhouse effect.

Results of the in-flight experiments were later published [6]. The solar wind detector followed the ups and downs of solar energy, including a spike on 15th December and scientists calculated that the solar wind took seven hours to reach Earth.

Venera 2 and 3 were the first spacecraft to fly under the new régime of Lavochkin, and the problems they encountered obliged Georgi Babakin to re-design. First, the domes and liquid-based radiators were removed. The radiators were moved from the solar panels to a ring behind the high-gain parabolic antenna. Instead of liquid, gas cooling was introduced and the high-gain antenna was adapted to serve as a radiator, taking away heat from the sunward side and dispersing it on the side facing away from the sun. As an additional precaution to maintain communications, the antenna would be turned continually toward Earth. Second, Lavochkin completed a new thermal testing chamber in January 1967, enabling testing of the equipment to be carried out to extremes of heat and vacuum. Third, Lavochkin made a 500 G centrifuge. When this first tested the existing 3MV landing cabins, they were quickly destroyed, which indicated that they must be reinforced if new landers were to have any chance of survival. Now the bottom and sides comprised new ablative material and three layers of absorption material.

FIRST DESCENT

Two Venus missions were attempted in June 1967. These spacecraft were heavier, at 1,106 kg. The first, Venera 4, got away successfully on 12th June. Five days later, a companion just made it into orbit – but no farther – and was renamed Cosmos 167. The block L upper stage failed to fire because the turbo-pump had not cooled before ignition, due in turn most likely to incorrect installation.

Even after the re-design, there were further improvements. The omni-directional antenna was replaced by a low-gain cone-shaped antenna extended on a bracket from the solar panels. Low-thrust engines were positioned on the astro-navigation system on the mid-course manoeuvre rocket. The cabin was designed to withstand 100 atmospheres, temperatures of 11,000° and stresses of 300 G. It could even float if it came down in an ocean, though few thought that likely and the movement detector experiment was taken off at this stage. Venera 4 stood 3.5 m high, its solar panels spanned 4 m and had an area of 2.5 m² and the high-gain antenna was 2.3 m in diameter. The descent cabin weighed 383 kg, comprising instruments, altimeter, thermal control, battery and two transmitters in a pressure shell, topped by a lid and parachute. Venera 4 not only carried the same gas analysis set first installed on Zond 1, but also a hydrometer to measure the amount of water vapour in the atmosphere. Telemetry would be sent from two transmitters at 1 bit/sec on 922 MHz. A small red flag was drilled into the side of the cabin and the letters USSR (CCCP in cyrillic) painted in red on the side. A radar was designed to spring open at the moment of parachute deployment and send back readings from the height of 26 km down to the surface.

Venera 4 in assembly

The parachute was designed to open as soon as a pressure of 0.6 atmospheres was sensed. Some scientists, still disciples of Tikhov, expected a watery surface and Venera 4 contained a sugar lock designed to release a transmitter should the craft splash down.

Venera 4 carried out a course correction on 29th July when 12m km from Earth so as to hit Venus on 18th October, otherwise the probe would have missed the planet by 60,000 km. The 338m km journey took 128 days, and 115 communication sessions were held. For the in-flight phase, a 2 m long magnetometer took readings. An ultraviolet spectrometer was built by Vladimir Kurt (b. 1932) to detect gases on the way to and around Venus.

This time, communications held up and signals were received as the probe closed in on the planet. The large dish at Jodrell Bank was following the mission and noted the cessation of signals. Then, as the pre-dawn darkness cleared, listeners at the radio telescope heard fresh, different signals coming from Venus. They presumed at once that they were listening to signals coming from the surface of the planet Venus.

Venera 4 cabin instruments

Thermometer, barometer	Vera Mikhnevich
Radio altimeter	
Gas analyzers	Kiril Florensky

Venera 4 mother craft instruments

Magnetometer	Shmaia Dolginov
Cosmic ray detector	Sergei Vernov
Ion detector	Konstantin Gringauz
Spectrometer	Vladimir Kurt

In reality, what was happening was more complex. The mother craft had indeed been put on an accurate trajectory for Venus and had released its descent craft, 44,800 km out over the nightside of the planet. Temperatures rose to 11,000°C. Its speed was broken by the upper layers of the atmosphere, subjecting the cabin at one stage to up to 450 times the force of gravity (450 G) and, after it slowed to 1,032 km/hr, the cabin released its 2.2 m^2 drogue parachute and then its 55 m^2 diameter main parachute with thick lines. The capsule swung underneath its canopy as it descended through the mist from an altitude of 52 km. Instruments began to send back data on temperature, pressure and the chemical composition of the atmosphere (gas analyzer). Temperature control kept the cabin at −8°C. The descent point was latitude 19°N, longitude 38°N in *Eisila* region.

The first temperature recorded was 39°C, pressure less than one atmosphere, but both soon began to rise. The parachute straps were designed to take a temperature of up to 450°C. The pressure built up as Venera 4 descended, 10, then 20, then 22 atmospheres. Twenty-three sets of readings were returned from the instruments. After 93 min, the cabin cracked open, probably at its weakest section at the top. Venera 4 had descended more than 26 km through the fog, but when the mission ended the

Venera 4 cabin

capsule was still far above the surface, at around 25–27 km. Here at 22 atmospheres, the temperature was 277°C and the atmosphere was largely carbon dioxide, with 1% oxygen.

Originally, Jodrell Bank thought these were surface signals, rather than telemetry radioed during the descent. Initially, this was the Russian view as well, for the final temperature and pressure readings were very much what they had expected at the surface. As if to confirm this, the altimeter was supposed to activate at the 26-km point and the aerodynamic data suggested that the probe had descended 26 km. Therefore, it must be on the surface.

In the event, the altimeter had not been accurately calibrated – it was intrinsically difficult to set it for another world – and had come on at 52 km. At one stage of interpreting the data, the Russians toyed with the idea that Venera 4 might have landed on top of a very high mountain! We now know that the cabin reached an altitude of around 26 km above the surface, maybe even getting down to 22 km, different histories giving slightly different final heights.

The descent of Venera 4 presented problems for Soviet news management, which really disliked uncertainty. Possibly still feeling sore about Jodrell Bank 'stealing' the Luna 9 pictures the previous year, a Russian statement immediately refuted Jodrell Bank's opinion that the spacecraft had reached the surface [7]. Then the space ministry decided to formally announce that a landing on Venus *had* been achieved after all. This official position became less and less tenable as Soviet scientists re-analyzed the data – and then compared them with fresh American information. The Americans launched a single probe to Venus during this window, Mariner 5. This small probe, 245 kg, flew past Venus at a distance of 3,991 km, much closer than its Mariner 2 predecessor, measuring the nature and composition of the planet's atmosphere and ionosphere. Although Mariner 5 found a temperature of 267°C, a closer look indicated that this temperature was well above the surface. Projecting its data, higher temperatures and pressures were indicated. Over the next two years, there was a number of meetings of Soviet and American scientists – in Tucson (Arizona), Kiev and Tokyo – and between them they were able to synthesize the Venera 4 and Mariner 5 data to make a coherent picture [8]. Eventually, they formed the view that Venera 4 had been destroyed long before reaching the surface. The Soviet press revised its announcement of a landing, realizing that even a descent into such a hostile planet was not something to be embarrassed about.

Science from Venera 4

Carbon dioxide	90–95%
Nitrogen	7%
Molecular oxygen	0.4–0.8%
Water vapour	0.1–1.6%
Temperature	270–280°C at point of crush
Pressure	20 kg/cm^2 or
	15–22 atmospheres at point of crush

No radiation belts, magnetic fields found

The science return from Venera 4 was greater than any previous Soviet Venus mission. In its interplanetary flight, Venera 4 found an increase in solar radiation levels compared with Venera 2 and 3, explicable by the rising stage of the solar cycle. Venus had no radiation belts and no magnetic field was detected, so if there was one, it must be less than 3/10,000ths that of Earth. Using Vladimir Kurt's ultraviolet spectrometer, a weak hydrogen corona was detected over Venus – 1/1000th that of the Earth – and the ionosphere was weaker than Earth's, 9,900 km over the nightside. Some weak atomic hydrogen was indicated 10,000 km out. This was evidence of water having leaked out of Venus's atmosphere a long time ago. The hydrometer found the atmosphere to be almost dry – an impossible finding granted its thick clouds! The carbon dioxide result was immediately contested. Venus experts had expected the ratio of carbon dioxide and nitrogen to be the other way around and American scientists were quick to suggest that the findings must be in error. Overall, the findings had the proponents of a watery Venus in headlong retreat and this was the last probe with a sugar lock.

Some Russians continued to hope that they had actually made a soft-landing with Venera 4, finding it hard to believe that surface pressures and temperatures could still be higher. But, over the following months, they came to accept that Venera 4 had indeed been crushed in an atmosphere that had proved to be much thicker than what they had anticipated. They had greatly underestimated the density of the atmosphere and the time it would take for a spacecraft to reach the surface. The Lavochkin bureau had quickly found itself on a steep learning curve. The next cabin would have to be

Mikhail Marov

stronger, tougher and come down more quickly if it were to have a chance of transmitting from the surface. A faster descent could be achieved by smaller parachute shrouds, but the outcome would be a more rapid landing, requiring a higher level of shock resistance.

Leading up the communication of the results was Mikhail Marov (b. 1933), later to become one of the communicators of the Russian interplanetary programme. Mikhail Marov spent four years in OKB-1 from 1958, before moving to the Department of Planetary Physics and Aeronomy in the Keldysh Institute of Applied Mathematics. He published a model of Venus, based on Venera 4's results, the first of hundreds of papers on Soviet Venus and Martian probes. He was personally responsible for numerous experiments on subsequent probes.

VENERA 5, 6

Positively, Venera 4 had done everything expected of it and the failure to reach the surface was hardly the probe's fault. For the first time ever, a Soviet interplanetary probe had not suffered an equipment failure. For the next launch window, in January 1969, two 3MV-type spacecraft were prepared, though Babakin's engineers called them the V-69 mission. For the first time, both left Earth successfully, on 5th and 10th January respectively, the first 100% success launch rate. The burn out of Earth orbit took 228 sec (Venera 5). Uncorrected trajectories would give a miss of 25,000 km for Venera 5 and 150,000 km for Venera 6. Accordingly, corrections were made on 14th March by Venera 5 when it was 15.5m km out, the change in velocity being 9.2 m/sec. Venera 6 adjusted its course on 16th March, making a necessarily larger velocity change of 37.4 m/sec when 15.7m km out. The transit time for both probes was 130 days. By the time they reached Venus on 17th and 18th May, Venera 5 had held 73 communications sessions and Venera 6, 63.

Venera 5 descent

Important modifications had been made to the Venera. Opinions are divided as to whether the V-69 missions were a genuine attempt to reach the surface, based on revised but still optimistic predictions of surface temperatures and pressures; or whether a deeper descent into the atmosphere, giving better data, was the best that could be hoped for. The cabin was strengthened, its temperature tolerance limit raised to 320°C, its pressure limit raised from 18 atmospheres to double that and its tolerance of G forces raised from 300 G to 450 G. The parachute was much smaller, only 15 m across. Both Venera 5 and 6 weighed 1,130 kg, their descent cabins weighing 405 kg. An improved altimeter was fitted, so that there could be greater certainty about whether the probe reached the surface or not and how close it got. A photometer was added to the instrument suite. Although the descents were scheduled for night-time, Venus had long been observed to have a certain level of nighttime glow which, it was hoped, would be measured.

Venera 5's descent cabin separated at 37,000 km distant from the planet, entering the atmosphere at 11.18 km/sec with an approach angle of 65°, a stress level 50% higher than Venera 4. The atmosphere slowed the probe down to 210 m/sec, at which point the small parachute opened, the descent began and transmissions to Earth commenced. The probe strained, groaned and eventually broke up when pressures rose above 27 atmospheres and temperatures above 320°C. Inside the cabin, temperatures had risen from 13°C to 28°C before yielding to pressure. Signals relayed a read-out from all the instruments every 45 sec. Venera 5 penetrated 36 km into the cloud, getting down to around between 26 km and 16 km altitude and had transmitted for 53 min. The landing point was latitude 3°S, longitude 18°, east of *Navka Planitia*. During the descent, the first attempt was made to gauge the light level, the photometer taking a single reading just 4 min before destruction and finding that there was a light level of 250 W/m^2. Full chemical sampling took place at 0.6 and 5 atmospheres.

The next day, Venera 6 separated at 25,000 km and plunged into the atmosphere, descending 37.8 km into the clouds and transmitting for 51 min before being likewise crushed. The Venera 6 landing point was latitude 5°S, longitude 23° east of *Navka Planitia*, some 300 km distant. The Venera 6 photometer failed, so the Venera 5 reading could not be validated. But Venera 6 seems to have got much farther down, possibly between 10 km and 12 km over the surface. The atmosphere was sampled at 2 and 10 atmospheres. Between the two, they were able to pinpoint the precise proportions of gases in the atmosphere, all the more important in the context of Venera 4's contested findings.

Once again, the Russians had hoped that they had got surface signals, or else mountain top signals. Venera 5 and 6 confirmed that their probes had still only penetrated the atmosphere and had yet to reach the surface. Publicly, the Soviet media played down expectations that surface transmissions would be achieved, indicating that a landing might be attempted the next time. The scientific data were similar to Venera 4. Again, the probes had functioned as best they could under the circumstances. For the Russians, the success of Venera 5 and 6 had come at a good time, for it took place the week before the Americans sent Apollo 10 around the moon.

By this stage, the debate on life on Venus was coming to a close. At the time of Venera 4, it was accepted that animal life must now be ruled out and that the

Venera 6

carboniferous swamps had probably boiled off. Some primitive, carbon-dioxide-breathing forms of vegetable life might still persist [9]. Now *Pravda* ran an article in which the anonymous chief designer of the Venera probes admitted that the planet was not very suitable for life and it was unlikely that humans could ever land there. A few die-hards clung to the idea that some silicon-based bacteria could still hang on, even in such hostile conditions [10].

Science from Venera 5, 6

Carbon dioxide	93–97%
Inert gases	2–5%
Nitrogen, oxygen	Less than 0.4%
Water vapour content	Traces of 4–11 µg/litre
Temperature	327°C (at 18 km, Venera 5); 294°C (at 22 km, Venera 6)
Pressure	27.5 atmospheres (Venera 5), 19.8 (Venera 6)

Descent transmission times

Venera 4	93 min
Venera 5	53 min
Venera 6	51 min

NEW TRACKING SYSTEMS

Even as Venera 5 and 6 sped toward Venus, the tracking fleet out to sea had been expanded and developed. The original tracking fleet had been built to support the 1960 1MV Mars probes and the crucial burn out of parking orbit. Since then the expansion had been driven by the need for tracking facilities for the lunar and manned programme, but they continued to have an important role for the interplanetary programme.

The original tracking ships had been the *Illichevsk*, *Krasnodar* and *Dolinsk*. Since then, they had been joined by a second group of converted merchant ships, the *Ristna* and *Bezhitsa*. Later, there was a third generation of four ships: the *Borovichi*, *Kegostrov*, *Morzhovets* and *Nevel*. Typically, these were ships in the order of 6,100 tonne displacement and a crew of 36.

A number of dedicated tracking ships was then commissioned. The first large tracking ship was the *Cosmonaut Vladimir Komarov* (17,000 tonnes), which appeared in the English Channel in summer 1967, to be followed by the flagship *Cosmonaut Yuri Gagarin* (45,000 tonnes), then the *Academician Sergei Korolev* (21,250 tonnes) and, much later, the *Academician Nikolai Pilyugin*. These were big, white, futuristic, sleek ships with giant domes and aerials, veritable homes at sea for their crews who would spend months at sea at a time. Smaller purpose-built tracking ships, in the order of 9,000 tonne displacement, joined in the early 1970s: the *Pavel Belyayev*, *Georgi Dobrovolsky*, *Viktor Patsayev* and *Vladislav Volkov*.

Back on land, new tracking stations were commissioned. The deep space net-work was to comprise three stations. First, there was the Yevpatoria set of stations, already described, the western part of the deep space tracking system, the TsDUC. The second station to come on line was Ussurisk, at the other end of the Soviet Union, where a 70 m dish was built, the eastern part. Then, finally, a third station was built near Moscow, called *Medvezhyi Ozera*, or Bear's Lake. Originally Bear's Lake was introduced for the Indian satellite *Bhaskira 2*, but its main function was for the

Tracking ship *Cosmonaut Yuri Gagarin*

interplanetary programme. This was a new 64 m receiving station, the height of a 15-floor house. In the air, Illyshin-18 propellor aircraft were converted to pick up telemetry from ascending rockets. Called the Illyushin-20RT (*ReTranslyator*, or Relay Station), four such aircraft were based in Baikonour from 1972 onward [11].

The position of receding deep space satellites was calculated by the Command and Measurement Complex. This comprised, in its initial form, 15 scientific measurement points (in Russian, NIPs), positioned across the land territory of the Soviet Union. Originally it was formed to track the first R-7 launch in August 1957, but was subsequently expanded, as part of the military, into a broad range of space-tracking functions using a mixture of radio, radar and optical tracking. The data gathered were fed into a large computing and coordination centre run by the Academy of Sciences. For interplanetary spacecraft leaving Earth, the optical parts of the tracking system would be of most value and the main optical tracking station was Zvenigorod near Moscow, supplemented by astronomical observatories in the Crimea (radio astronomy) and the Caucasus (visual) [12].

Mars, Venus spacecraft series

1MV	1960–1
2MV	1962
3MV	1963–72 (Venus) 1964 (Mars)
M69	Mars 1969
M71S, M71P	Mars 1971 (orbiter, lander)
M73S, M73P	Mars 1971 (orbiter, lander)
4V1	1975–8 (Venus)
4V1M	Venera 13, 14 (Venus)
4V2	Venera 15, 16 (Venus)
5VK	VEGA (1984)
UMVL	1977–96 (Phobos, Mars 96)

Large tracking dish

UNDER NEW MANAGEMENT

The transfer of the Soviet interplanetary programme from Korolev's OKB-1 to Lavochkin's OKB-301 began to show results after two years with the successful descent of Venera 4 into the atmosphere of Venus. The double success of Venera 5 and 6 showed that this was not just a lucky outcome, but a radically improving record in reliability. Even if the launchers continued imperfectly, the actual spacecraft were reaching their destinations and performing their missions to the limits of their designs. The real test for Lavochkin lay ahead, with the development of its first indigenous probes destined for Mars and fitting them to Vladimir Chelomei's Proton rocket, still untested for planetary missions. Would this combination be more successful than the earlier series and the Molniya?

In the meantime, Sergei Korolov passed from the scene. Taken in for a routine operation on 5th January 1966, he expected to be back in good time to celebrate his birthday on the 14th. The operation went horribly wrong, complications developed and he died on the operating table. The organizational genius who developed the Soviet space programme and who inspired and begun the interplanetary programme was no more. The chief designer was given an enormous funeral, the biggest since that of Stalin. The funeral eulogies spoke of the devastating effect of his death and the enormous loss that his departure inflicted on the space programme. Their rhetoric might have been considered an excessive genuflexion for the man who had brought a

Sergei Korolev died in 1966

devastated country to the cutting edge of cosmic exploration. But they did not exaggerate.

REFERENCES

[1] Tyulin, Georgi: Memoirs, in John Rhea (ed.): *Roads to space – an oral history of the Soviet space programme*. McGraw-Hill, London, 1995; for a broader description of the changes at this time, see Siddiqi, Assif: *The challenge to Apollo*. NASA, Washington DC, 2000.

[2] Harford, Jim: *Korolev*. John Wiley & Sons, New York, 1996.

[3] Ivanovsky, Oleg: Memoir, in John Rhea (ed.): *Roads to space – an oral history of the Soviet space programme*. McGraw-Hill, London, 1995.

[4] Mitchell, Don P.:
  – Remote scientific sensors;
  – Biographies at *http://www.mentallandscape.com*

[5] Huntress, W.T., Moroz, V.I. and Shevalev, I.L.: Lunar and robotic exploration missions in the 20th century. *Space Science Review*, vol. 107, 2003.

[6] Gringauz, K.I., Bezrukih, V.V. and Mustatov, L.S.: Solar wind observations with the aid of the interplanetary station Venera 3. *Kosmicheski Issledovanya*, vol. 5, #2, Nauka, Moscow, 1967, as translated by NASA Goddard Space Flight Centre, 1967.

[7] MacPherson, Angus: Venus – on target. *Daily Mail*, 19th October 1967.

[8] Mitchell, Don P.: *Plumbing the atmosphere of Venus*, at *http://www.mentallandscape.com*

[9] MacPherson, Angus: Venus – on target. *Daily Mail*, 19th October 1967.

[10] Moscow cheers landing on Venus. *The Observer*, 18th May 1969.

[11] Gordon, Yefim and Komissarov, Dmitry: *Illyshin-18, -20, 022 – a versatile turboprop transport*. Midland Counties Publication, Hinckley, UK, 2004.

[12] Smid, Henk: Soviet space command and control. *Journal of the British Interplanetary Society*, vol. 44, #11, November 1991.

5

First landfall on Venus, Mars

There was a time when the idea of learning the composition of celestial bodies was considered senseless. That time has now gone. To observe Mars from a distance of several dozen miles, to land on its satellites and even on the surface of Mars – what could be considered more extravagant? But with the advent of reactive devices a new and great era in astronomy will begin.

– Konstantin Tsiolkovsky: *Collected works*. Moscow, 1956.

FIRST LANDFALL ON VENUS

The descents of Venera 4, 5 and 6 had showed just how dense and difficult was the Venus atmosphere. Another Venus window came around in August 1970. Once again, measures were taken to strengthen the descent cabin for the two probes to be launched. Whatever the purpose of the Venera 5 and 6 missions, this new set was intended from the start to reach the surface intact, even if that meant it was overbuilt and that hardly any scientific instruments could be carried.

The parachutes were made ever smaller. They included a number of improvements over Venera 5 and 6. Submarine designers were consulted, so as to ensure that the spacecraft hull could survive under intense pressures. This probably made the cabin much heavier, at 490 kg. New materials, such as titanium, were used. The cabin was constructed as a perfect sphere, with no holes, welds or sub-structures, with the instruments under the top hatch, which would blow off to release the parachute. The shell was then reinforced with shock-absorbing material, so as to withstand stronger pressures (between 100 and 180 atmospheres) and temperatures (530°C) and thus ensure a hard-landing. A 90 min surface transmission time was aimed for. The cabin was to be chilled to −8°C before atmospheric entry. In ground tests, the descent cabins were put into the thermal vacuum chamber, pressurized to 150 atmospheres, cooked to 540°C and the chamber flooded with carbon dioxide – as close to Venus on Earth as you could get.

The designers tried to reconcile the contradictory requirements of as rapid as possible a descent with as soft as possible a touchdown. To do so, the parachute had a double system of cords. To speed the initial descent, the parachute was only partially opened under nylon cords. The glass nitron parachute strings were designed to be held tightly together by cords and then, as the cords melted, the parachute would fully open to slow the final descent. In another break from previous practice, the lander would not deploy far above the atmosphere. Instead, the mother craft would stay attached until the two broke apart high in the Venusian atmosphere – but it would benefit from its coolant until the last possible moment.

Two 3MV series spacecraft were prepared for the 1970 launch window (in Lavochkin, the V-70 missions). Venera 7 entered parking orbit of 182–202 km, 51.7°, and after a burn of 244 sec left Earth orbit at 07:59 on 17th August and by 10 a.m. Earth was receding 40,000 km in the distance. The attempt to launch a second

Venera 7

probe showed that the problems with the 8K78M upper stage had still not been fully overcome, for the fourth stage again failed on 22nd August. The payload was left stranded in a 208×890 km orbit, $51.1°$. Due to some form of transformer failure, block L ignited later than planned and cut off after 25 sec instead of the 244 sec planned. Attitude control was lost and by the time block L passed over Europe an orbit later, it was tumbling wildly [1]. The failure was given the cover name of Cosmos 359 and the spacecraft crashed back to Earth on 6th November.

Venera 7 instruments
Mother craft
Cosmic ray detector.
Solar wind detector.

Lander
Thermometer.
Pressure meter.

The first course correction for Venera 7, made on 2nd October 17m km out, does not appear to have been sufficient, and on 17th November, 31m km out, a second was made, the first time two burns had been made on a Venera mission. *En route* to Venus, Venera 7 picked up the signals of a powerful solar flare on 10th December 1970, one which was also noted by the Lunokhod observatory on the moon. Venera 7 was due to arrive at Venus on 15th December after its 120-day journey. By the time it did so, 124 communications sessions had been held, Venus was now 61m km away and signals took 3 min 22 sec to travel that distance. Three days before encounter, the landing cabin was activated and its batteries charged.

This is what should have happened. The spacecraft came in at 11 km/sec or 724 km/hr, at an angle of almost $70°$ to the horizontal, experiencing 350 G, with its heatshield reaching $11,000°C$ temperatures. Entry would take place on the night side of the planet some 2,000 km from the sunny zone. The intended landing point was latitude $5°S$, longitude $351°$, east of *Navka Planitia*. The parachute was to open at 60 km over the imagined surface, triggered by a detector as soon as it recorded a pressure of 0.7 atmospheres at the beginning of the cloud layer.

All day, people waited for an announcement from Moscow. None came and the worst was feared. Three days later, Radio Moscow reported that signals had been received for 35 min during the descent and that the cabin had then failed. Obviously, the additional changes to the descent cabin were not enough.

Instruments during the descent had recorded temperature and pressure. At parachute deployment, temperature was $25°C$ and pressure 0.6 atmospheres. All was well. Thirteen minutes later, the cords melted through and the main parachute came open. Six minutes later, the descent began to go badly wrong: the parachute tore and the cabin began swinging wildly and a few minutes later collapsed altogether, leaving the cabin to free-fall the rest of the way, with temperatures recording $325°C$ and pressures of 27 atmospheres. Impact speed was 17 m/sec, slightly

Venera 7 top antenna

less than 60 km/hr. The signals faded, came back full on for 1 sec and then vanished altogether.

Attempts to listen in to disappeared spacecraft rarely had happy endings, as the efforts to find Beagle 2 on Mars reminded us many years later. To make things worse, the transmitter during the descent had interrogated and sent back readings from only the temperature sensor and the rest of the descent data had been lost (although it was later possible to infer atmospheric pressure).

There things rested until a month later, after the new year break. Oleg Rzhiga (b. 1930), a radar astronomer and an expert in signal processing, decided to re-run the tapes of the descent signals. It must have been a tedious and ear-straining job to listen in to the cosmic crackle, with all the squeaks, loose air, beeps, static and pips and feed them into the reading machine. It was all what analysts call 'noise'. But, amazingly, there it was, barely discernible and extraordinarily weak, a signal from the surface of Venus! The fresh signal had come in soon after the main signal had cut out. Subsequent analysis of the data suggested a sudden acceleration in the rate of descent, which had been recalculated to last 60 min. It was originally thought that Venera 7 hit a sudden downward air pocket, causing a fast descent, disruption to the signal and a fast landing. More recent analysis suggests that the parachute detached at 3 km, with the probe free-falling the rest of the way, making it lucky that the cabin survived at all [2]. In fact, the signal on landing was interrupted at the very start, suggesting that the cabin bounced, the full signal coming back on for 1 sec at the top of the bounce.

For 23 min, Venera 7 had indeed transmitted from the surface of the planet Venus, from *Navka Planitia* at latitude 5°S, 351° longitude. The temperatures were even, meaning that the probe could no longer be moving. The data collection system for

Oleg Rzhiga

recording pressure failed, but the pressure value was later imputed from other measurements, being constant at 92 atmospheres, five times greater density than estimated in 1967. Temperature was 475°C. Scientists calculated a surface wind of 2.5 m/sec. The 495 kg probe had sat there for over 20 min, cooking away in a temperature sufficient to melt lead or zinc, before succumbing. But they had done it, after all. Almost ten years after Venera 1 had been launched, a Soviet planetary probe had completed its mission and soft-landed on another planet. Whatever the many misfortunes that had affected other aspects of the descent, Venera 7 showed that a landing could be accomplished. At that moment, the Lunokhod rover was driving across the Sea of Rains on the moon, so the Soviet Union was receiving transmissions from two heavenly bodies at the same time.

The reason for the weak signal was that when it landed – or beforehand – the transmitter was jolted out of alignment, by about 50°. It seems that Venera 7 landed hard and came to rest on its side. Although Earth was overhead, the signal was pointing to one side, so only about 1% to 3% of the normal strength was received.

Venera parachute canopy size

Venera 4	$50\,m^2$
Venera 5, 6	$15\,m^2$
Venera 7, 8	$2.5\,m^2$

BASELINE VENUS: VENERA 8

A repeat was organized for the next window, in spring 1972, called the V-72 missions. Venera 8 was launched on 27th March 1972, after a 243 sec burn out of Earth orbit. Venera 8's onboard rocket made a course correction on 6th April.

Venera 8

Venera 7 had shown that a soft-landing could be achieved. Now that the design was known to be sound, it was possible to install a sophisticated suite of instruments on the next probe so as to radically improve the level of scientific knowledge obtained, both during the descent and after landing. Venera 7 had been designed to withstand a surface pressure of 180 atmospheres, much more than what was necessary now that the surface pressure was known to be 90 atmospheres. Accordingly, Venera 8 was designed around a pressure limit of 105 atmospheres, the weight saved going for thermal protection, scientific instruments and a stronger parachute. The new model was then tested in wind tunnels and in a model Venus atmosphere, including the carbon dioxide.

Venera 8 carried a light measurement indicator, ammonia detector, speed indicator and gamma ray spectrometer to test the soil. Because of new instrumentation, the lander weight was up again, to 495 kg. For the first time, Venera 8 was designed to come down in daylight, with instruments to measure brightness, so as to pave the way for later spacecraft with cameras.

Venera 8 instruments: mother spacecraft
Solar wind detector.
Cosmic ray detector.
Ultraviolet spectrometer.

Lander
Temperature and pressure sensors.
Anemometer.
Photometer.
Gamma ray spectrometer.
Gas analyzer.
Altimeter.

This time, instruments were carried to test the nature of Venus's soil. These had been developed by Valeri Barsukov (1928–1992) in the Vernadsky Institute of Geochemistry in Moscow, the body which had already supervised the investigation of lunar rock brought back by Luna 16.

As was the case the previous time, the companion failed (Cosmos 482). This was, incidentally, the last time a 8K78M failed to send a spacecraft to another planet and also marked the end of the use of the 8K78M launcher for the interplanetary programme and the end of the 3MV series. Although there were occasional failures of the 8K78M on other missions after 1972, these were infrequent. Due to a timer failure, block L fired for only 125 sec, providing a velocity increase of 1,498 m, instead of 3.4 km/sec, leaving Cosmos 482 in an extreme orbit of 205 × 9,805 km, which decayed after six years. This probe would have landed on the dark side, so as to compare Venera 8's findings on the sunlit side. In the event, it appears that in July 1972 the pre-programmed lander did indeed separate from the main spacecraft and make a descent – but into the Earth's atmosphere! [3].

While leaving Earth, Venera 8 achieved an escape velocity of 41,433 km/hr. After another 117 days, Venera 8 approached the planet, aimed at the heart of the then crescent-shaped sunny side of Venus, a slim early morning crescent low over the horizon at Yevpatoria 108m km away. The batteries of the descent craft were charged 1.1m km out and a refrigeration system fanned cold air at −15°C around the cabin so that the instruments could survive longer. Separation took place at an approach speed of 41,696 km/hr some 53 min before atmospheric entry. Venera 8 came in a 77° at 11 km/sec. Too steep an angle and the craft would burn up; too shallow and it would bounce back into space.

By the time Venera 8 had decelerated in the upper atmosphere and reached 67 km, it was down to a speed of 250 m/sec. The parachute was reefed some 60 km above the surface, to be then fully opened at 30 km altitude. This time the parachute was only 2.5 m across, a dramatic reduction compared with the 50 m of Venera 4. A radar in the H-shape of a television aerial was deployed over the side of the descent craft, scanning downward. The landing target was latitude 10°S, longitude 335°, east of *Navka Planitia*. The landing zone was less than 500 km in diameter and had to be both on the daylight side and in line of sight with Yevpatoria.

At 50 km, the instrumentation turned on, including photometers, temperature and pressure sensors, ammonia detector and radar altimeter. The first radar reading came in at 45 km, the last at 900 m and 35 data points were received. The capsule was blown 60 km sideways during its 53 min descent, or slightly more than 1 km/min (over 60 km/hr). The radar was able to generate a profile of the ground over which

Venera 8 in preparation

the cabin was travelling, mapping two mountains (one 1,000 m, another 2,000 m), a downward slope, a hollow 2,000 m deep and then a gentle slope upward to the landing spot.

After a 53 min descent, Venera 8 thumped down on the planet, cutting the parachute lines free as it did so. There it transmitted for 63 min. On top of the tub-shaped lander were a cone-shaped main antenna, light intensity meter, ammonia detector and temperature and pressure sensors. This time, the spacecraft carried an omni-directional antenna to make sure there would be no problem of signal strength or direction. Earth was 30° over the horizon at the time. The antenna was set on tripod pads and ejected a couple of metres from the spacecraft on a snaking cable. It was designed to survive heavy winds and to function in the event of the lander being blown over or turned upside down. The omni-directional antenna had a footnote, for one of the causes of the loss of the Beagle 2 spacecraft on Mars in 2003 was considered to be that the aerial was pointing the wrong way – a lesson the Russians had learned in 1970.

In any case, the main spacecraft antenna was functioning properly. First data came in a 13 min stream from the primary antenna and included temperature, pressure and light levels. There was then a 20 min stream from the omni-directional antenna and a 30 min stream from the main antenna. Temperatures on Venus differed little from the daytime and nighttime side, they found. Venera 8 carried a gamma ray spectrometer, which made measurements during the descent and took two readings on the surface itself, signalled back to Earth within 42 min of landing. This gave a reading of potassium 4%, 2 ppm uranium and 6.5 ppm thorium, indicating alkaline basalt. Subsequent geological analysis suggested that the Venera had come down in a zone of granite, flowing lava and small volcanoes. The soil had a density of 1.5 g/cm^3. The radar altimeter had the ability, during the descent, to estimate the density of the surface layer, calculated at 800 g/m^3, which would make it like granite. The landing area was upland plain, reckoned to be typical of 65% of the planet. The finding of granite-like rock by Venera 8's instruments was contested, as most of the surface is basaltic, but subsequent mapping shows that the instruments were correct.

The light levels told a lot. Venera 8's photometer took several measurements during the descent and on landing. Venera 8 seemed to come out of the cloud layer at 35 km, even though most of the cloud layer is about 10 km higher. Venera 5 had given back just one flash of light from its nighttime descent, but Venera 8's descent was in the daytime. Light was measured at 27 points during the descent. Light decreased steadily from 50 km to 35 km as the spacecraft went through the cloud layers. Then, from 35 km downward, out of the cloud layer, the level decreased only slowly and would have been like Earth on a cloudy day. The photometer measured the surface brightness as poor, similar to Earth's at sunrise, giving visibility to the distance of about 1 km. Only 1% of sunlight reached the surface – a finding that had to be treated cautiously, for the Sun was only 5° over the horizon at the time. The atmosphere below 32 km was essentially transparent. And, on the question of the clouds, the gas analyzer gave the first hints to the mystery of the dry clouds, finding sulphuric acid. Now we know that the upper clouds are a fine sulphuric haze.

The scientific haul from Venera 8 provided the baseline knowledge of the planet, with confirmed figures for temperature, pressure and atmospheric composition [4]. For the first time, measurements were made of the nature and composition of the rock. Venera 8 profiled the cloud layer, finding that it was more like a fog layer and that it had two main levels: thick cloud at the top layer (65 km to 50 km), then haze between 35 km and 48 km, but clear below this level. The cloud – or fog – had high winds at parts of the upper level, but the winds were more gentle on the surface.

The Venus baseline: science from Venera 8

Temperature	470°C.
Pressure	93 atmospheres.
Soil	Loosely composed, uncompacted, made of alkaline basalt: 4% potassium, 0.002% uranium and 0.000 65% thorium.
Wind	100 m/sec above 48 km, between 40 m/sec and 70 m/sec between 42 km and 48 km, but only 1 m/sec for the last 10 km down to the touchdown point. At 11 km altitude, winds followed the direction of planetary rotation. Surface wind at 1 m/sec.
Atmospheric composition	97% carbon dioxide, 2% nitrogen, 0.9% water vapour, 0.15% oxygen. Ammonia of 0.01 to 0.1% between 32 km and 44 km.
Surface density	1.5 gm/cm^3.
Cloud layer	Base of cloud layer: 35 km, with thin haze to 48 km and then a thick layer to 65 km. Skies more like fog than Earthly cloud.

Venera 8 represented the limit of what could be achieved by the 3MV series of Venus probes. At around the same time, the Americans brought their small Mariner spacecraft series to a successful conclusion. Mariner 10 flew in the next Venus window. The 408 kg spacecraft passed 5,800 km from Venus in February 1974, before becoming the first spacecraft to fly past and photograph Mercury. Three hours before reaching Venus, Mariner 10 turned on its infrared camera and for the next eight days filmed its rotating, swirling bands of clouds.

Mariner 10 demonstrated what could be done by the imaginative use of interplanetary trajectories. After its first close approach of Mercury at 680 km, it refined its course for three subsequent encounters with Mercury over the next two years. Mariner 10 mapped most of the planet, measured its temperature, found its magnetic field and detected minute traces of an atmosphere.

With Venera 8 and after ten years of efforts, the Russians had actually completed a soft-landing and surface transmissions from another planet, one which had proved immeasurably more difficult to explore than anyone had imagined. The development of more advanced spacecraft, based on the Mars designs from 1966, was approved in the case of Venus on 24th March 1973 [5]. But what progress had been made with the Mars designs meantime?

BABAKIN'S FIRST DESIGN: MARS 69

Following the disappointment with Zond 2, coupled with the findings of Mariner 4 that Mars' atmosphere was very thin, Georgi Babakin had decided not to use the 3MV series any more for Mars missions. Instead, using a new version of the UR-500 Proton rocket, he planned to develop a new, much larger spacecraft for Mars and then Venus missions. The development period for the new Mars probe was necessarily lengthy, meaning that the 1967 Mars window would be missed and that the first

Georgi Babakin, chief designer

realistic target date would be spring 1969. The Mars 69 missions (this was the title given them in the Lavochkin design bureau) were the first of this new generation of Mars and Venus probes. Subsequent probes in the series were identified by their year (M-69, M-71 and M-73). To save time, it was decided to model the M-69 design on the Ye-8 series of moon probes then in design for orbiting, rover and sample return missions.

The UR-500 Proton rocket opened up new possibilities for the designers of the Babakin bureau, enabling them to consider large probes up to five tonnes in weight. The original two-stage UR-500 Proton rocket could put up to 12 tonnes into low-Earth orbit. The rocket was then developed in two versions: a three-stage model, able to put 20 tonnes into low-Earth orbit and subsequently used for space stations (8K82K, or Proton K); and a four-stage version, able to send over four tonnes to geostationery, lunar or interplanetary targets (11S824 or Proton K-D).

Glushko's OKB-456 developed the first three stages, but the fourth stage that would be needed for interplanetary (or lunar) missions came from Korolev's design bureau. Called block D, it was developed in-house in OKB-1 by Mikhail Melnikov. This relied on the traditional liquid oxygen and kerosene that was the hallmark of Korolev's approach. The engine used, the 11D58, was a derivative of block L used on the 8K78M. This stage was small in size, 3.7 m in diameter and 5.7 m long and had the crucial role of carrying out the vital burn out of Earth parking orbit to the interplanetary target. It had a thrust of over eight tonnes, a specific impulse of 349 sec. Several versions of block D were developed, the variants for interplanetary missions being called the D or 11S824 (Mars 2–7, Venera 9–10) , D-1 or 11S824M (Venera 11–16, VEGA) and D-2 or 11S 824F (Phobos, Mars 96 [6]). The new Proton, with the block D, was first tested on Cosmos 146 on 10th March 1967, so it would be available for the spring 1969 launch window.

The launching window in 1969 was not a very favourable one and the spacecraft weight was limited to 3,495 kg, most of the balance being fuel. According to the preliminary design approved by Georgi Babakin in November 1967, it was intended to orbit Mars and send down a descent probe to profile the atmosphere, but a soft-landing would not be attempted. As it approached Mars, the spacecraft would release a cabin to enter the atmosphere at an angle of 10° to 20°. A parachute would be deployed at Mach 3.5. A wide margin was given for the time of parachute deployment, between as high as 31.7 km altitude and as low as 2.2 km altitude, with a transmission time before the surface was reached of between 230 and 900 sec. The descent craft would provide a scientific picture of the atmosphere so that, based on its findings, a proper soft-landing could be attempted in the following Mars window, spring 1971.

The infrared spectrometer for Mars 69 was developed by Vasili Moroz (1932–2004). He had begun studying Mars in 1954 from Alma Ata Observatory and would have been there during the time of the great Gavril Tikhov. Later he moved to the Institute of Space Research where he became head of planetary investigation. Using infrared spectrometry, he had become ever more convinced of the low atmospheric pressure on Mars and wanted to use such instruments to map Mars' pressure and temperature profile in detail.

As for the mother craft or orbiter, it would apply a braking speed of 1,750 m/sec and enter a Martian orbit with a periaxis of 2,000 km and apoaxis of between 13,000 and 120,000 km, with a period of rotation between 8.5 and 12 hours and an inclination of 35° to 55°. Here, it would map the surface of Mars, conduct remote sensing and identify landing sites for future missions. The orbital flight plan included the first

Vasili Moroz

interceptions of the two Martian moons, Phobos and Deimos [7]. An important aspect of the mission was to give future mission planners a much improved knowledge of Mars's orbit, path and movement through the sky – called *ephemeris* by astronomers.

DESIGN CHALLENGES OF MARS 69

As the design progressed over the winter of 1967–8, major difficulties were encountered. The fuel tanks proved to be especially problematical on a range of fronts, such as the exclusion of bubbles, sealings, membranes and calculating the centre of gravity as fuel was expended. The designers were aware that ensuring an accurate trajectory for Mars arrival was essential for the subsequent fast sequence of events: separation, entry, parachute opening and landing.

The original design shows two dumbbell-shaped fuel tanks, but many designers began to have doubts about how well the fuel tanks and engine systems would work, as well as the centre of gravity, balance and stability of the system [8]. Accordingly, Georgi Babakin scrapped the Ye-8 based preliminary design in February 1968, only 13 months before the opening of the Mars window and asked his designers to start again with a new, simpler tank design – one large spherical tank. There were

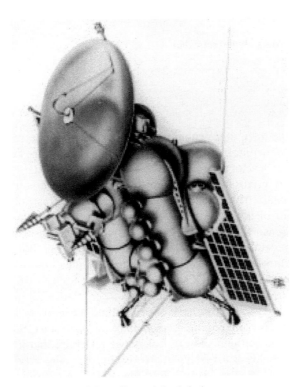

Mars 69 – original design

some important changes in the mission profile. Because of uncertainties about the trajectory, it was decided that the spacecraft should enter Mars orbit first, between 1,000 km (much lower) and with a apoaxis of 70,000 km and a period of rotation of 65 hours. The spacecraft's weight had now gone up to 3,834 kg, including a lander of 260 kg. The scientific instruments weighed 100 kg, including 15 kg on the lander. The following instruments were installed:

Instrumentation for Mars 69 orbiter
Magnetometer.
Meteorite detector.
Low-frequency radiation detector.
Charged particle detector.
Cosmic ray and radiation belt detector.
Spectrometer for low-energy ions.
Radiometer.
Multichannel gamma spectrometer.
Mass spectrometer for hydrogen and helium.
X-ray spectrometer.
X-ray photometer.
Ultraviolet spectrometer.
Infrared Fourier spectrometer.
Three cameras of 35 mm, 50 mm and 150 mm.

Instrumentation for Mars 69 descent cabin
Gas analyzer.
Pressure detector.
Atmospheric density detector.
Temperature detector.

NPO Guskov in the town of Zelenograd developed a 11 kg compressed data processing system for the scientific instruments, the most advanced of the kind available. The control system was developed by Stanislav Kulikov, much later to be head of the design bureau itself. Gyroscopes were developed by the Scientific Research Institute for Applied Mechanics of Academician Viktor Kuznetsov. Instead of the combined sun and star sensor of the 1MV series, this time there were no fewer than four sun sensors, two star sensors, one Earth sensor and one Mars sensor, developed by the Central Design Bureau for Geophysics and the Arsenal Design Bureau in Leningrad. For orientation, Mars 69 had nine helium-pressurized tanks supplying nitrogen gas to eight thrusters – two for pitch, two for yaw and four for roll. Here, high-pressure nitrogen engines were developed by the Central Scientific Research Institute of Fuel Automatics in Leningrad. This was part of the automobile industry and was chosen on the basis that – since car engines on Earth were required to start tens of thousands of times – they would have the necessary experience to ensure that an attitude control system near Mars could restart frequently too. The radiotelemetry system, weighing 212 kg, was developed by the Central Scientific Research Institute of Space Instruments of M.S. Ryazansky and comprised:

- three transponder receivers on 790–940 MHz, using 100 W of power;
- two 6 GHz impulse transmitters at 25,000 W operating at 6 kbyte/sec, with power 25 W, to send pictures;
- two 790–940-MHz transmitters at 128 bits/sec for telemetry, distance and velocity and commands;
- telemetry with 500 channels; and
- antennas: three semi-directional low-gain for in-flight, one 2.8 m high-gain for Mars arrival.

The camera system was designed to store 160 images of 1,024 pixels each side and had an onboard scanning system. There were three cameras, one 35 mm, one 50 mm for ground tracks of $1,500 \times 1,500$ km and one with a 350 mm lens for pictures 100×100 km. Several filters could be applied to the cameras: red, green and blue. The close-up system was expected to detect features on Mars as small as 200 m across. The solar panels had an area of 7 m^2, feeding 12 amp to a cadmium–nickel battery of 110 amp/hr storage.

The Lavochkin bureau drove at a fierce pace to get the two Mars probes ready in time. Engineers worked around the clock and it was not unusual to send around special cars to get managers out of their beds during the night to solve knotty problems. Other specialists brought folding beds into their workplace. The cafeteria was ordered to stay open 24 hours a day and provide free meals. There was no

Mars 69 – second design

Mars 69 completed

overtime pay. The head of the Ministry of General Machine Building, Sergei Afanasayev, worked hard to clear production problems from sub-contractors and other parts of the space industry. Some of the suppliers were scattered far away: the tanks were made in Orenburg (south of the Urals), the propulsion system in Ust-Katavisk and the landers in Omsk. Eventually all the parts came together over 1968, were assembled and tested. Assembly of the two spacecraft was declared completed just one hour before the final deadline of 31st December 1968. Final testing took place in January and the spacecraft were shipped to Baikonour in February.

Vibration-testing was a challenge for Lavochkin, for it required a level of violence and size far beyond the levels inflicted on their aircraft. Accordingly, the Mars probes were brought to another plant for this in Reutovo. Just as the test concluded, they thought successfully, a metal bracket gave way and the micro-engines fell off, one after another. A new, reinforced bracket was attached, vibration-testing recommenced and the test began anew – successful second time around.

But the spacecraft were vibration-tested without the lander, which was not ready. The designers were unable to keep the lander and its instrumentation within its weight limits and there was insufficient time to test the parachute system. At this late stage, the decision was taken to proceed with an orbiter-only mission. The final weight of the probe at launch was 4,850 kg.

The mission profile was revised again and was now to enter Mars orbit of 1,700 × 34,00km, 40°, period 24 hours – called the capture orbit. After several weeks, the engine would be fired to lower the orbit to 500 × 700km for three months of closer imaging and experimentation. Arrival dates were set for the 11th and 15th September 1969.

BROKEN WINDOWS, BROKEN HOPES

The Lavochkin teams accompanied their two Mars spacecraft to Baikonour in early February 1969. Part of the cosmodrome had been wrecked when the N-1 moon rocket had exploded on its first flight early that month. The N-1 blew out the windows of the hotel where the Lavochkin engineers were billeted and the central heating system, exposed to the −30°C steppe winds, duly froze. The windows were re-glazed but getting the frozen central heating system going again took longer and small electrical heaters had to be provided in the meantime, barely managing to keep room temperature above zero. The spacecraft were loaded onto the first Proton rockets built to carry probes to the planet.

The first Mars 69 failed on 27th March 1969, when there was a Proton third-stage rotor failure which caused a turbine fire at 438 sec. Signals ceased and the wreckage ended up in the Atlai mountains. Had it succeeded, it would have been called Mars 2, for it carried a commemorative medal on board – showing the solar system – marked 'Mars 2, 1969'.

When the second Proton took off on 2nd April, black smoke gushed immediately out of the right engine only 0.02 sec after liftoff. The other five engines continued to burn, but the huge rocket quickly lost direction and began to travel sideways rather than upward. The remaining engines were shut down and after 40 sec the first stage of the rocket exploded into a dense luminous black mass of fire only 3,000 m from the launch site. The second stage, amazingly, was still intact, lying on the ground, but liable to go up at any moment. Proton was going through the worst period of its difficult early history and many moon probes were lost during the same period. All the great efforts of Georgi Babakin and his engineers over the previous two years came to nought.

What was doubly galling was that the Americans went on to great success with Mars that year, launching Mariners 6 and 7. Barely days after Neil Armstrong, Michael Collins and Buzz Aldrin returned in triumph from the moon, the two Mariners reached Mars, flying past the snow-clad craters of the southern Martian polar regions, sending back low-resolution but nonetheless informative pictures. Mariner 6's 75 images and Mariner 7's 126 photographs mapped 9% of the surface, their remarkable achievement barely noticed in the euphoric, relieved and exhausted United States.

The problems with Proton not only claimed the Mars 69 probes as victims but a number of lunar missions over the next year. Eventually, Georgi Babakin persuaded Minister Sergei Afanasayev to introduce a requalification programme. This took place over spring and summer 1970, culminating in a suborbital test on 18th August

1970. This led to a radical improvement in performance, but Soviet space histories might have been happier ones had the changes been introduced sooner.

The disaster of 2nd April had an important consequence for the development of rocket technology in Russia. The launching attracted a high-level attendance of the military and what workers in the industry called 'the bosses' of party and government. The explosion sprayed toxic nitric fuels over a wide area. They were shocked at the contamination around the launch area and the fact that there was no procedure for cleaning it up: the only thing to do was wait for rain to wash the stuff away. Three years later a series of long-range design studies was commissioned for Russia's future rocket needs, called *Poisk*, and one of its first recommendations was not to use toxic fuels again, signalling a return to kerosene.

MARS 71S: AN ORBITER TO FLY AHEAD OF THE LANDING FLEET

The 1969 failures were a poor reward for the frantic efforts to build the spacecraft in time. Not only that, but they deprived the scientists in Lavochkin of the trajectory data and information they so badly needed to guarantee success for the soft-landings that they had planned for 1971. Georgi Babakin and his colleagues originally considered re-running the Mars 69 missions in 1971 for this purpose, with a view to postponing landings in 1973. The problem was that this seemed a waste of the uniquely favourable 1971 launch window, which was so good that it could permit combined orbiters and landers to fly with a large payload. By contrast, landings would be tough to achieve during the more difficult 1973 window – the missions would have to be split between orbiters and landers.

So it was decided to proceed with two soft-landers in 1971 but that an orbiter should fly ahead of them and give ground controllers sufficient information so that they could refine the entry corridor for the soft-landers following close on their heels. The two landers would have a pre-programmed system for entry to Mars, but ground controllers would update it with new data from the orbiter. A week or two before reaching the planet, the pre-programmed system would be updated as a result of the orbiter's new *ephemeris*, ensuring – it was hoped – a good entry angle and orbital insertion burn. The M-69 design was abandoned and a larger, newer design was introduced.

This strategy would also enable the Soviet Union to beat the Unites States into Mars orbit, but this – though important – was only a secondary consideration. The orbiter would be launched as early in the window as possible and given 800 kg extra fuel, double the normal load. The new plan was agreed at a meeting of mission planners convened by president of the Academy of Sciences, Mstislav Keldysh, in his original office in the Institute of Applied Mathematics in late May 1969.

The new orbiter was called Mars 71S (S for sputnik, or orbiter). The extra fuel required a further redesign of the tanking system, a task carried out by V.A. Asyushkin, subsequently designer of the Phobos spacecraft propulsion system. A third tank design, this time a cylindrical one, was introduced. Advantage was also taken of the opportunity of redesign to make changes to the system as a whole,

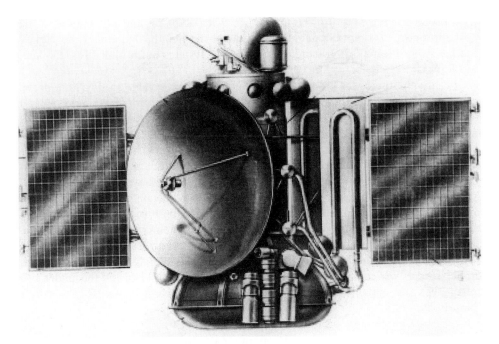

Mars 71S orbiter

especially to improve access to the spacecraft's instrumentation during ground-testing so that individual parts could be removed and reinstalled without major recabling. Weight was 4,549 kg, which included 2,385 kg of fuel and gas.

Mars 71S instruments
Imaging system.
Magnetometer.
Cosmic ray detector.
Two radiometers (one infrared).
Three photometers (one infrared, one ultraviolet).
Spectrometer.
Charged particle spectrometer.

For the Mars 71 missions, Babakin decided to transfer the development of the control system to the Scientific Production Association for Automatics and Instrument Development (NPO AP), the large company of Nikolai Pilyugin, one of the original designers in Korolev's council of designers dating to 1946. Nikolai Pilyugin, born 5th May 1908 (os) in Tsarskoye Selo, was the king of guidance and had learnt his trade in postwar Germany analyzing the guidance system of the German *Wasserfall* anti-aircraft missile, arguably a more sophisticated rocket than the V-2. The reason for going to Pilyugin was not clear, for a system had been devised in-house for the

M-69 probes. There were prolonged rows between Babakin and Pilyugin as to the most appropriate system, Pilyugin wanting the N-1 system he had already developed (too heavy, said Babakin), instead offering lighter systems but not till 1973 (too late, said Babakin).

Eventually, there was a compromise. Pilyugin's new system on the Mars probe would not only control the Mars probe but the Proton's fourth stage, block D and bring its weight down to 167 kg. He guaranteed it would be accurate enough for all manoeuvring, orbiting and the descent.

The orbiter-landers were two big spacecraft. These were called Mars 71P, P for *pasadka*, or lander. Each was a cylindrical doughnut 3 m tall (4.1 m with capsule on top), with a large dish antenna, omni-directional antenna and two silicon solar wings. A course correction rocket was fitted on the bottom. Radiators carried pipes to circulate inert gases. The lander was encapsulated on the top. On the outside were astro-navigation sensors, two cone-shaped low-gain antennas and magnetometers. On the back of the solar panels were diminishing cylindrical cones. They were antennas for the data link with the descent capsules. The final weight of the Mars orbiter-landers was set at 4,650 kg – the orbiter taking up 3,440 kg and the lander and cone 1,210 kg, both fuelled. The dry weight of the orbiter was 2,265 kg. The spacecraft comprised:

- propulsion system, bottom, with separate parts for nitric acid fuel and oxidizer, and gimballed engine;
- lander, placed in shell at the top;
- two solar arrays, 2.3 × 1.4 m, spanning 5.9 m;
- 2.5-m parabolic high-gain antenna;
- main antenna using 928.4 MHz and small antenna – for communication with lander – on the solar panels;
- navigation instruments: sensor to measure the planetary angle, Earth sensor, star sensor, precise solar sensor and rough solar sensor;
- experiments.

HOW TO REACH AND SOFT-LAND?

The Mars arrival was a crucial stage, for the spacecraft must be sure of its exact position, deploy the lander in the precise place for its descent and then fire its own engine for correct entry into Mars orbit. This was a crucial navigational test, one requiring extremely accurate knowledge of Mars's *ephemeris*. Although Soviet astronomers had good optical information for Mars, only the Americans had the experience of passing the planet with operational spacecraft. The OKB Lavochkin planners studied in detail the American reports from their Mariner 4, 6 and 7 missions. They asked informally for the unpublished *ephemeris* arising from these missions, but the Americans were not prepared to provide them [9].

The plan was that seven hours before Mars arrival, 50,000 km out – called the point of first measurement – the optical measurement system would orientate the spacecraft for Mars encounter. The lander, in its cone, would then be separated for six hours of independent flight. It would turn its nozzle into the path of flight and carry out a 100 m/sec braking manoeuvre and then begin its descent. For the mother craft, a 'second measurement' would take place at 20,000 km out. This would calculate the point at which the engines should fire to brake into Martian orbit, intended to be the point of closest approach to the planet, 2,350 km (± 1,000 km). The burn was scheduled to impart a velocity of 1,190 m/sec.

This was a different and more complex manoeuvre than that used many years later by the Americans and Europeans to land the Huygens probe on Titan. There, the entire assembly of Cassini–Hugyens, mother craft and lander, headed on a collision course for Titan. The mother craft dropped the lander and then made an avoidance manoeuvre. For these Soviet Mars missions, there would be a common orientation, but after separation the lander would have to make its own engine firing to enter the atmosphere while the mother craft would make its own orbital injection manoeuvres.

The designers spent many long hours considering the best way to get the lander down to the surface and how to cope with their uncertain knowledge of the density of the atmosphere. In the end, they resolved on a 3.2 m diameter cone with an angle of 120°. They went for the widest possible cone, maximum stability during entry, fastest braking and soonest opening of the parachute. A landing using only rocket engines was considered, but – in the light of insufficient knowledge of air density – was adjudged to have a low chance of success and a combined parachute and rocket system chosen instead (the Americans went through a similar set of options in deciding on the method for their Viking spacecraft). The parachute would come open at mach 3.5.

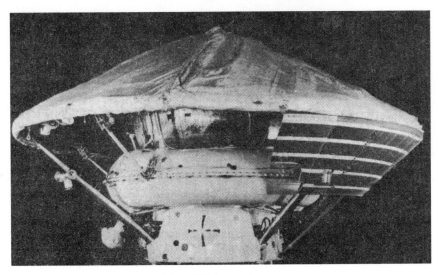

Mars 71P lander in cone

For the final touchdown more options were considered. The touchdown would be triggered by a radar altimeter which would provide the final engine firing and command a gunpowder rocket to pull the parachute free. Rubber bags were considered, for these had been used successfully for Luna 9 to soft-land on the moon. The designers felt that there was a probability though that they would be ruptured by the gases from the engines and that as a result they would burst. Instead, it was decided to provide strong 200 mm thick protective plastic foam around the lander to soften the final touchdown. Once on the surface, the four petals would open to right the craft and expose the instrumentation.

Thankfully, when it came to testing parachute system, the Scientific Research Institute of Parachute Landing Facilities (NII PDS) of N.A. Lobanov already had experience of high-speed parachute deployments. Lobanov built them a drogue chute of $13\,m^2$ and a main parachute of $140\,m^2$. A small balloon was built to lift a model lander 32 km high and then drop it for landing tests. Models were also lifted much higher, to 130 km, by the M-100B meteorological rocket. Fifteen such tests were carried out. The person who led the team for Mars landing aerodynamics was Yuri Koptev, who was later to head the Russian Space Agency from 1992 to 2004. Five model landers were made available for high-altitude tests and different dimensions of parachute diameter were tried. A catapult was also used to fire models of the lander at simulated Martian surfaces, from horizontal impact speeds of 28.5 m/sec to vertical speeds of 12 m/sec and the lander was rated to absorb impacts of up to 180 G. Full simulated landings were made. During one, the lander made a simulated touchdown in a vacuum facility, transmitted for 25 min, sent a rover on a short trip and for

Yuri Koptev

25 hours was subjected to a pressure of 6 mb and an airflow of 25 m/sec as well as extremes of temperature. The whole cycle, including a fresh communications session, was then repeated. The actual lander's parts were sterilized before launch so as to prevent contamination.

All this presupposed that the landers would survive the tricky entry into the Martian atmosphere. The descent module was a flattened cone 2.9 m in diameter. Attitude was controlled by gas microengines and four gunpowder engines. The entry sequence was marked by the following milestones, each designated T:

- Separation at 46,000 km distance from Mars.
- After 900 sec, firing of gunpowder engine at 120 m/sec to commence entry.
- After 1,250 sec, orientation for entry. Rotation using small nozzles at the edge of the cone.
- Entry to Mars atmosphere at 5,800 m/sec (100 sec), with 10 key points (T):
  T1   prompted by accelerometer, activation of drogue chute at mach 3.5;
  T2   after 10 sec, pulling out of furled main chute;
  T3   activation of instruments;
  T4   after 12.1 sec, full inflation of main parachute;
  T5   after 14 sec, release of cone;
  T6   after 19 sec, activation of landing altimeter – descent is now at max. 65 m/sec;
  T7   after 25 sec, soft-landing engine is pulled out from the top of the landing module and is now hanging under the parachute;
  T8   after 27 sec, low-level radar is activated and engine is ready to fire – this stage may last between 20 and 200 sec;
  T9   high-level radar detects surface within 16–30m and activates soft-landing engine;
  T10 low-level radar shows speed at less than 6.5 m/sec and commands soft-landing engine to lift parachute free.

After touchdown, which must be at a speed no greater than 12 m/sec, there would be a delay of 15 sec for the cabin to settle. After 15 sec, a command would open the aeroshell cover, pushed open by compressed air. The instruments would then be activated. Data were to be transmitted at 72,000 bits/sec on two channels, with panoramic images of 500 × 6,000 pixels. Each communication session would be 18–23 min. Signals would be sent for three or four days.

The lander comprised a egg-shaped module weighing 358 kg and measuring 1.2 m across. Its objectives were to:

- soft-land on the planet;
- return images of its surface with two television cameras with 360° view;
- provide data on meteorological conditions with a windgauge, thermometer and pressure gauge;
- test the mechanical and chemical properties of the soil; and
- provide information on atmospheric composition (mass spectrometer).

Mars 71P lander

The experiments weighed 16 kg. The lander had a mass spectrometer to study the atmospheric composition, temperature, pressure and wind; a scoop to test the surface; and four aerials to communicate with the orbiter. Batteries provided several days of power supply and there was a system for thermal control. Instruments varied a little between one probe and the other.

MINI-ROVERS

The landers carried a small walking robot or skid rover, called PrOP-M, with a mass of 4.5 kg and tethered to the craft for communications. PrOP-M stood for *Pribori Otchenki Prokhodimosti–Mars*, literally 'instrument for evaluating cross-country movement'). These were developed by Alexander Kemurdzhian (1921–2003), the tank designer whose VNII Transmash company built the Lunokhod moon rovers. The skid rover was a squat box 250 × 200 × 40 mm with a dynamic penetrometer and radiation

Alexander Kemurdzhian

densitometer. PrOP-M was designed to walk on skis up to 15 m, the limit of the cable. It was programmed to stop to make measurements every 1.5 m. The skid rover had built-in artificial intelligence: when it met an obstacle, it was programme to reverse and use the skids on alternating sides to walk around the obstacle. Their existence was not acknowledged until April 1992.

As for the surface pictures, the landers carried pinpoint photometer cameras of the type used for the first soft-landing on the moon by Luna 9. Although often described as a television camera, it was more accurately called a pinpoint photometer and took the form of a cylinder with a space for the scanning mirror to look out the side. These are optical–mechanical cycloramic cameras and do not use film in the normal sense, instead scanning for light levels, returning the different levels of light by signal to Earth in a video, analogue or digital manner.

The mass spectrometer was built by atmospheric scientist Vadim Istomin (1929–2000). He had flown his spectrometers into Earth's atmosphere (on sounding rockets) and were so sensitive that they had detected the ions of vaporized meteorites there.

Mars 71S and the second lander were each equipped with a French experiment, the first time foreign equipment had been carried on an interplanetary mission. Called Stereo, it was designed to study solar radiation at 169 MHz and took the form of a T-shaped aerial installed on the solar panel. The old habits of secrecy died hard, though, for the French simply handed the experiment over: not only were they not involved in integrating their experiments with the spacecraft, but they were not

Vadim Istomin

even permitted to see drawings of how this was done [10]. The French were simply told it had been duly installed, but were not told where – in practice, one set was on Mars 71S and the other on Mars 3.

As for the orbiters, their mission objectives were:

- image Mars's surface and clouds;
- determine the physical and chemical properties of the surface of the planet;
- measure the properties of the atmosphere;
- monitor solar wind, interplanetary and Martian magnetic fields; and
- act as a communications relay for the landers.

The mother craft to circle Mars had a large suite of instrumentation, broadly similar for each craft. The weight of instruments on Mars 3 was 89 kg. Mars 2 and 3 marked the first use of the Zufar (350 mm) and Vega (52 mm) cameras, designed by Arnold Selivanov at the Institute of Space Device Engineering. Each weighed about 9 kg. An improvement on those planned on Mars 69, they carried enough film to take 480 pictures. Pictures would be taken in sets of twelve, each one 35–40 sec apart. It was intended that most of the imaging be done during the first 40 days in Mars orbit during the points of closest approach. An inventory of the instrumentation on Mars 71S is not available, but it is known to have been equipped with at least a radiometer and magnetometer.

Mars 71P system of photo-relay

Mars 71P orbiter experiments
Infrared radiometer in the 8–40 µm range to study temperature – able to measure down to
     −100°C.
Photometer to analyze water vapour concentrations.
Infrared photometer.
Ultraviolet photometer to detect hydrogen, oxygen, argon.
Lyman-α sensor to detect hydrogen in atmosphere.
Visible range photometer to study reflectivity of surface and atmosphere.
Radio telescope operating on a 3.4-cm radiometer, 60 cm tall, to determine reflectivity of surface
     and atmosphere, giving temperatures just below the surface (50 cm).
Infrared spectrometer to measure carbon dioxide.
Photo-television unit with one narrow-range 350-mm camera (Zulfar) and one 52-mm wide-
     angle camera (Vega) with red, green, blue and ultraviolet colour filters using an onboard
     scanning system of 1,000 lines and 10 m resolution.
Radio occultation experiment for information on structure of atmosphere.
Spectrometer to find water vapour and carbon dioxide.
Radiometer to determine temperature of the surface.
Magnetometer.

In-flight experiments
Instrument to measure galactic cosmic rays and solar corpuscular radiation.
Eight electrostatic plasma sensors to determine speed, temperature and composition of solar
     wind in 30,000–10,000 eV.
Three-axis magnetometer to measure interplanetary, Martian field on boom from solar panel.
Fluxgate magnetometer.
Cosmic ray particle detector.
Electron and photon charged particle spectrometer.

Lander
Mass spectrometer.
Temperature and pressure sensors.
Anemometer.
Cameras.
Soil analyzer.
Skid rovers PrOP-M.

THE LOSS OF MARS 71S AND ITS CONSEQUENCES

The Soviet Union of course had the benefit of knowing the American schedule for
Mars probes for 1971. To follow their success with Mariners 6 and 7 in 1969, the
Americans planned to send more Mariners, 8 and 9, this time to orbit the planet.
Although Russia's orbiter would be launched a day after the launch date set for
Mariner 8 – that was the way the windows operated from the two different launch
sites – the Russian spaceship would still overtake Mariner 8 and get there first.
     The pressure on the USSR relented, for Mariner 8 crashed into the Atlantic Ocean
on 9th May 1971. The Mars 71S Soviet spacecraft, the orbiter, went ahead the next

day, a Proton rocket putting a Mars probe in Earth parking orbit for the first time (145 × 159 km). An hour and a half later, Mars 71S should have fired over the Gulf of Guinea out of Earth orbit, but it transpired that an incorrect eight-figure code had been sent to the spacecraft for block D ignition. It was renamed Cosmos 419 and fell out of orbit after two days. For Lavochkin, this was a dispiriting start to the new generation of Mars probes. Proton had now let them down three times out of three. In the midst of this, the French had lost their Stereo experiment, but were told nothing.

The loss of the Mars 71S orbiter had serious practical consequences for the loss of navigational and trajectory data and meant that the Russians would not have a good *ephemeris*. They could no longer update their pre-programmed entry and orbit burn manoeuvre commands from a fresh *ephemeris*. Instead, the two probes would have to make precise optical measurements of their position close to the planet when they arrived and overwrite the pre-programmed commands in real time. This was a complex, risky exercise, more sophisticated than anything they had attempted before.

ON THE WAY AT LAST

Now it was time for the Mars 71P series. Mars 2 entered Earth parking orbit of 137 × 173 km, 87.44 min, 51.8° on 19th May. This time, the burn went perfectly. When Mars 2 was already 2.48m km from Earth, Mars 3 followed nine days later on 28th May and it was straight away announced that soft-landings were intended.

Although it was stated that everything was working perfectly on Mars 2 (in fact everything was), the French were puzzled that they were getting no data from their solar radiation instruments. This obliged the Russians to explain that there was an electronic fault with them and, sadly, data could not now be returned. Years later, the French found out that they had been placed on the now-destroyed Cosmos 419. America's Mariner 9 went up on the last day of the launch window, 30th May. This time, the USSR had been outsmarted by the Americans, for its faster trajectory guaranteed that it would get to Mars first. Unless Mariner 9 failed, Russia would not be the first country to put a spacecraft in orbit around Mars.

By the end of July, Mars 2 was 17m km away from Earth, Mars 3 following 2m km behind. With Earth receding in the distance, the probes recorded the Earth's magnetic tail stretching out a distance of 19.2m km. Mars 2 used its rocket motor to change its course on 17th June and Mars 3 the same on 8th June. Throughout the summer, Radio Moscow issued progress reports on the two spacecraft speeding away from Earth at a rate of 192,000 km every 24 hours. Communications sessions were held almost every day – in practice, every night because the spacecraft had line of sight with Yevpatoria only at nighttime. Things seemed to be going well at last.

Mars 3 carried on its solar panels the second set of the French figure-of-eight 1 m Stereo experiment to measure radio emissions from the sun. This was the first of the in-flight experiments to give results and, at the end of August, Stereo recorded hydrogen particles in the solar wind travelling at between 300 km/sec and 600 km/sec.

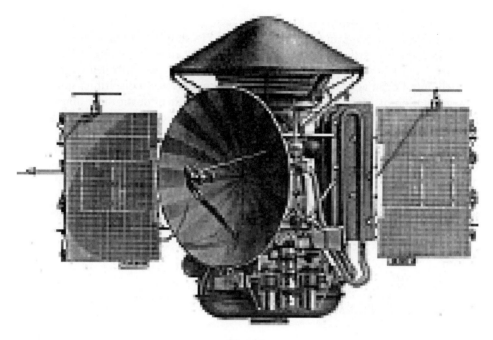

Mars 2

GEORGI BABAKIN, 1914–1971

In fact, although this was not mentioned at the time in the upbeat reports of Radio Moscow, the ground controllers had already overcome a serious problem. A crisis developed on 25th June. The decimeter transmitters on both spacecraft failed at almost exactly the same time. Each switched to the backup transmitters, but they in turn quickly failed too. The centimeter band transmitter was activated and, although it responded, it became apparent that it was pointing away from Earth. Identical triple failures on two spacecraft simultaneously was quite outside the normal run of expectations and left ground control utterly baffled. Lavochkin staff flew out to Yevpatoria, being obliged to use military aircraft because all civil airliners were packed with happy vacationers *en route* to the Black Sea. Once there, they tried to reconstruct the circumstances which led to the failure of communications. They managed to get the original transmitters to re-broadcast, but at a reduced rate. They never recovered the centimeter band transmitter and took the precaution of always shielding these antenna from the sun thereafter.

On 3rd August, the chief designer of the institute, Georgi Babakin, died of a heart attack. He was only 57 and by all accounts at the height of his creative powers, already planning new missions [11]. He had guided Lavochkin through the six years that followed OKB-301 being given responsibility for interplanetary missions. It was not just rhetoric to call him a worthy successor of Korolev in the field of deep space missions. He had seen the successful soft-landing on Venus by Venera 7 and two of his

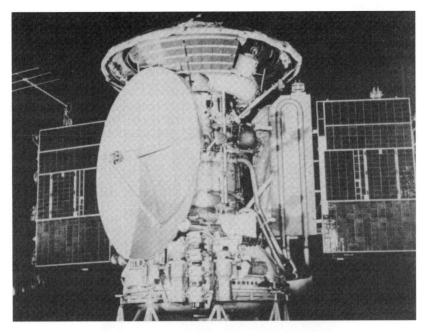

Mars 3

new probes were at last on their way to their destinations. The scientific research unit in Lavochkin was named in his memory.

In his place was chosen Sergei Kryukov, who had been with Babakin's bureau for only a year. Ten years earlier, in 1961, Korolev had appointed him a deputy chief designer, along with Vasili Mishin and a number of others, so he was well regarded. Siddiqi [12] calls him a tall, bespectacled, quiet man. Sergei Kryukov was born 10th August 1918 in Bakhchisarai in the Crimea, his father being a sailor and his mother a nurse. His mother was ill throughout his early years and died when he was eight. Young Sergei spent much of his childhood in an orphanage, but relatives eventually removed him and ensured he got an education. He caught up quickly, entered Stalingrad Mechanical Institute in 1936 and then its artillery facility, continuing to work there as the city was under German siege. With the war over, he continued his education in the Moscow Higher Technical Institute, while getting work in the #88 artillery institute there. No sooner had he started than he was transferred to Germany, his task being to reverse-engineer Germany's advanced guided missile, the *Schmetterling*. On his return, he transferred to work for Sergei Korolev in OKB-1, where he developed the R-3, R-5 and R-7 rockets, being number 4 in the design of the R-7 after Korolev, Tikhonravov and Mishin. His contribution was recognized by an Order of Lenin. After the R-7, Kryukov went on to work on upper stages, principally Molniya's block I and block L. Assigned to develop block D for the Proton and the N-1, he fell out with Vasily Mishin in 1970 and managed to transfer to NPO Lavochkin, never imaging that within a year he would become director.

Sergei Kryukov

As if the death of Babakin were not enough, ominous signs appeared for the upcoming planned landings. The American Mariner 9 began to test its cameras at the end of October. They brought worrying news. Instead of a cold, quiet, southern spring Mars, the planet was convulsed in a raging dust and sand storm. It was covered in swirling red and brown clouds. Not a thing could be seen. Mars 2 and 3 were pre-programmed and would land in the middle of it. Next, Mariner 9 claimed an unusual planetary 'first' for the United States – the first probe to orbit another planet, for it entered Mars orbit on 14th November. The cloud storm did not present a serious problem for Mariner 9, which simply waited out the storm and went on to carry out one of the most successful Mars-mapping exercises of all time.

FIRST TO REACH THE SURFACE OF MARS

Final course corrections were made by Mars 2 and 3 as they approached the storm-engulfed planet. Mars 2 corrected its path on 20th November and made a final adjustment on the day of arrival, the 27th. Mars 3 corrected its path on its day of arrival, the 2nd December.

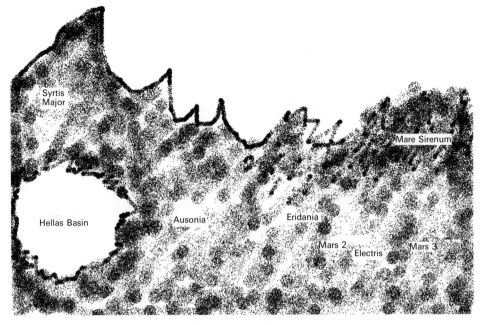

Mars 2, 3 landing sites

Mars 2 duly arrived on 27th November 1971 after 192 days in transit. Optical orientation took place at the appointed moment seven hours before landing, with a view to ensuring the correct angle of descent for the lander and the right point of retrofire for the mother ship to enter Mars orbit. The navigational and automatic systems, never tried before, worked perfectly, targeting the lander correctly for entry and landing. Mars 2 was on a perfect path for detaching the lander and burning into orbit. But, at this stage there was an electronic failure and tragically the new orientations were not transferred to the main computer [13]. The operations for landing and orbit entry proceeded anyway, but according to the old, pre-programmed trajectory calculated the previous May, not the new, updated on-the-spot orientation. The lander was detached. Its motor fired briefly to slow the craft for Mars entry. After four hours, the cone entered the Martian atmosphere at 21,600 km/hr. Shock waves formed, heating it to several thousand degrees. Because the trajectory could not be corrected, it came in at far too steep an angle and hit the Martian surface a couple of minutes before the parachute could even be deployed. Mars 2 smashed to bits before the lander could transmit information. We can be certain that the lander, with Soviet pennants on board, became the first Earth object to reach the surface of Mars at 44.2°S, 213°W in a rounded hollow called *Eridania*, 480 km east of the *Hellas* Basin.

As for the orbiter, it fired its engine but at the point calculated the previous May. The consequences were much less serious and did not affect the ability of the probe to enter Mars orbit, although it was not quite the orbit intended.

MARS 3: FIRST TO SOFT-LAND

Mars 3 was travelling one million kilometres behind. On 2nd December, after a journey of 188 days and 70,000 km from Mars, it orientated itself for the forthcoming set of complex manoeuvres. Again, the process worked perfectly. Mars 3 used its optical navigation system to orientate itself properly. But. this time the navigational data were correctly transmitted for the separation and subsequent trajectory of the lander.

Separation took place 4.5 hours before entry. The cabin was spun to obtain stability and then its rocket fired briefly to set the cone for entry. Now it was de-spun, so that the parachute would not tangle later. Mars 3 began its descent in the southern hemisphere in the light-coloured regions called *Electris* and *Phaetonis*. It took only 3 min for Mars 3 to go through the atmosphere and for the parachute to open. An accelerometer triggered a pilot parachute and then a reefed main parachute, still at supersonic speed. At mach 1 the main parachute was inflated. The heatshield, glowing red hot, was dropped and the radio transmitter started to send data. The altimeter turned on. The descending cabin must have been buffeted by the ferocious winds then blowing across the dusty red landscape. Twenty-five metres above the surface, a tiny rocket above the cabin blasted the parachute free so that it would not fall on top of the 450 kg landing cabin, and another rocket fired briefly to slow the final landing speed. This was a sophisticated combination of manoeuvres designed to minimize the shock of descent and landing, while keeping the weight down.

The soft-landing was achieved at 16:50.35 on 2nd December 1971. Four petals opened and the domed shape of the capsule rested there on the sands of Mars, the first-ever soft-landing. Antennae popped out, aerials searched skywards and TV cameras

The edge of Mars from Mars 3

began scanning. Video transmission began 90 sec later at 16:52.05 on two independent channels at 72 kbyte/sec (kbps) – and then, exasperatingly, fizzled out after only 14.5 sec.

Here began one of the mysteries of the space age. At the time, no one knew why the link was broken. Moscow Radio announced that 'no discernible difference' was received in the video, but did not release it in any case. Some suggested a transmission failure, others that the mother ship failed, others more sinisterly that no message had ever been received and this was a hoax. American interpretation of the data indicate a fairly hard landing at 20.7 m/sec which may not have helped. Radio Moscow stated that the cabin was designed to withstand not only hard-landings but surface winds of up to 360 km/hr. The site was 44.9°S, 160.08°W not far from the northern limit of the south polar cap.

Ground control did not become aware of the situation until some time later. The landers did not transmit directly to Earth, but through the orbiters. Here, there was nearly a fresh disaster. The orbiters were required to stabilize themselves half an hour after orbital injection, using their sun sensor. The reason for the delay was to give the soot of the recently-fired engine time to disperse and not dirty the fragile optics of the sun sensor. A few minutes after it was opened, the sun sensor failed, presumably as a result of dirt and the spacecraft lost stabilization. Wisely, there was a backup sun sensor, commanded to turn on an hour later; and this time the soot had dispersed, because the lock was maintained. Ground control commanded the spacecraft to relay

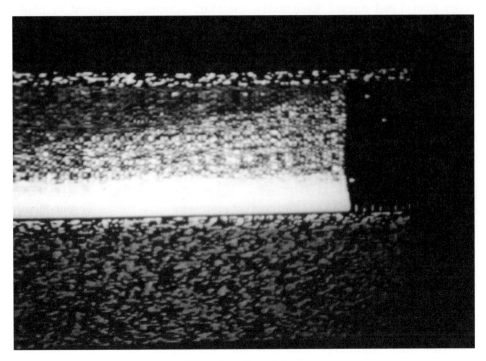

Mars 3 surface picture

Mars 3 horizon, enhancement

the tape of the surface transmission. In all the replays, the transmissions fizzled out after 14.5 sec. To add to the frustration, they were at a loss to understand how two independent transmitters could fail simultaneously. Later, designer Vladimir Perminov learned of how – during the Second World War – dust storms combined with coronal discharges could have strange effects on radio transmissions and maybe that is what happened.

Years afterwards, going through the Moscow archives of the Vernadsky Institute in the days of *glasnost*, some researchers were amazed to find . . . the first picture from Mars. The signal was printed out again and there it was – the picture of a flat horizon and rocks in the foreground! The photo would not have won a competition, but it was proof that a photograph had indeed been received. Apparently at the time the Soviet censors had decided that it was of too poor quality to release and that no picture was better than a bad picture. Even today, the Mars 3 picture is disputed by photo analysts. Some dispute that the picture does actually show a horizon and photo enhancement has not been successful in obtaining a satisfactory level of further detail. Others insist that the image was actually upside down because the wind had blown the probe over. What's clear, though, is that Mars 3 did soft-land, did transmit some form of signal from the surface and that its mission was then ended prematurely, probably by the dust storm.

THE ORBITAL MISSIONS

Despite the failure of the Mars 2 landing, the breakdown of Mars 3's signals and incorrect orbital insertions, the two mother craft went on a successful nine-month scientific mission that did not conclude until September 1972. These have often been overlooked in the accounts of the Mars 2–3 missions. The return was in fact quite substantial [14]. The instruments of Mars 2 and 3 examined the environment around Mars (its atmosphere in general and dust storms in particular) and the surface of the planet, especially its thermal profile. The outcomes were all the greater an achievement for – with the loss of the centimeter band transmitter – the return of data from Mars orbit was likely to be limited. Thankfully, radio designer Ryazansky came to the rescue, for he found a way of sharply increasing the data rate transmission from the decimeter band transmitters.

Mars 2 had entered a close orbit – 1,380 × 25,000 km – a lower apoaxis than

Mars crescent from Mars 3

intended, circling the planet every 18 hours (25 hours had been planned) at 48.9° inclination. Mars 3 had a similar periaxis (1,500 km) but flew much farther out – 190,000 km – taking 12 days 19 hours to complete an orbit at 60°. Some estimate that Mars 3's orbit may have extended out even farther, to 209,000 km [15]. Mars 3's orbit was so far out as to compromise its scientific return. The reasons for the far-out orbit are not clear. It is possible that the data for the correct entry burn was not transferred to the computer at the point of orientation, even though it was successfully done so for the lander and that the engine fired according to the pre-programmed information. The other, simpler possibility is that the burn was too short [16].

In fact, the quite different orbits may have yielded a richer scientific outcome than might otherwise have been expected. Communications sessions took place daily. Most of the information was transmitted from December 1971 to March 1972, though signals continued until 22nd August 1972, the formal end date for the missions, by which time Mars 2 had completed 362 orbits and Mars 3 twenty. All this time, Mars's distance from Earth was increasing and by the end was over 230m km away. Mars 2 experienced serious telemetric problems and it is no coincidence that most of the scientific data are referenced to Mars 3.

As was the case with their lunar orbiters, Soviet Mars orbiters were built to carry out their imaging mission soon after arrival in orbit. There were several main reasons for this. First, remembering the Luna 9 experience when Jodrell Bank had scooped their moon pictures first, they wanted to be the first to ensure the arrival of their own data themselves. Second, Russian spaceships had much shorter design lives than American spacecraft of the same period, so missions should be accomplished speedily. Imaging was a top priority, so it should be done first, especially when maximum solar

Mars 3 picture of a distant Mars

power was available. Although it had been intended to take most of the photographs during the first 40 days in orbit, this had to be postponed because of the dust storm, and the first pictures were not shown on Moscow television until 22nd January.

Photography of Mars was impeded by two factors. First, the sand storm engulfing Mars meant that the surface of the planet was invisible until well into 1972. Second, the cameras were keyed on too light a setting, giving their pictures an over-exposed appearance. As was the case with Mariner 9, photographic data began to improve from January 1972 when the dust storm abated. The first pictures showed cratered regions, mountains up to 15 km high and craterless plains (presumably the *Hellas* Basin). Pictures showed mountains of up to 3,100 m and depressions of 1,200 m below the reference line. They picked up volcanic peaks up to 22 km above the datum line (the massive *Olympus Mons* is now estimated at 24 km).

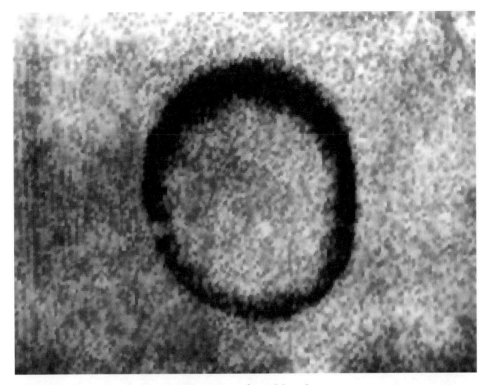

Mars crater from Mars 3

Both the wide-angle and narrow-focus cameras used filters to get measurements in different parts of the spectrum. Mars 2 concentrated on close-range imaging, while Mars 3 took wide-angle pictures from a distance and was in a good position to observe the planet-wide dust storm. Mars 2 and 3 sent back 60 pictures (most being Mars 3), film being scanned on board at 1,000 lines and then transmitted back to Earth electronically, like that of the Automatic Interplanetary Station in 1959 and Zond 3 in 1965. Mars 3 took some striking pictures of the thin crescent of Mars beckoning in the distance from the apoaxis of its orbit. The normal procedure was to turn the main scanning instrumentation on for about 30 min at the point of closest approach to the planet. For example, on Mars 3 the radiotelescope was switched on during the 1,500–5,000-km phase of its elongated orbit and covered $200\,km^2$ at a time. The two Mars spacecraft concentrated on the regions from 30°N to 60°S. Mars 3's main batches of photographs were taken in four sessions: on 10th and 12th December, 28th February and 12th March. One of the first pictures showed the limb of the atmosphere with the dust swirling within the thin air. Of the two transmitters to be used by Mars 3 to send pictures, one – the high-power impulse transmitter – failed, so only low-resolution versions at 250 lines were sent. The radio telescope was developed by radio astronomer Arkady Kuzmin (b. 1923), who had used instruments for years to measure planetary atmospheres, pressures, densities, composition and temperatures.

Mars mountains from Mars 3

An early discovery was that of orbital anomalies, apparent by the first week of January 1972, akin to lunar mascons and an observable flattening of the Martian poles in the order of 35 km. Mars 2's perigee was bent from 1,250 km to 1,100 km in one month and by April the scale of the anomaly was measured to average 150 km.

The probes charted the interaction between the Martian atmosphere and the solar wind. Mars appeared to form a shock wave to resist the solar wind, for the probes entered and exited from areas of charged particles. The French Stereo was turned off on 25th February after providing 1 MB of data over a 185 hr observation, a sizeable amount in those days [17]. The magnetometer suggested a very weak magnetic field, but a more likely interpretation of the data is that it was really a distortion of the interplanetary magnetic field and not a true field caused by Mars's core.

On 16th February, Mars 2 flew over *Hellespont*, *Iapygia* and *Syrtis Major*, marking elevations up to 12 km and 15 km high, measuring pressure (5.5–6 mb) and temperatures. Mars 3 identified water [18]. Cloud particles were detected up to an altitude of 40 km. The proportion of moisture or water vapour was tiny, 'only the thickness of a human hair', with a level of water concentrations 1/2,000th to 1/5,000th that of Earth, and most of it was found in equatorial regions. If all the water in the atmosphere were to be condensed, it would cover the surface to a mere 15 microns. There was still sufficient atmosphere to present a perceptible airglow 200 km behind the terminator at night. High in the atmosphere, long, thin clouds were found in the ultraviolet filter. At higher altitudes of 100 km, the carbon dioxide atmosphere tended to break up into carbon monoxide and oxygen. Traces of oxygen could be found as high as 800 km and a hydrogen corona was found 10,000–20,000 km above the planet.

Mars 2 and 3 found that the dust storms reached up to 10 km high in the atmosphere. In their report on the great dust storm of 1971, Soviet scientists concluded that:

- the storm lasted three months and was highly irregular in nature, being intense in some parts of the planet and less so in others;
- the particles, being mainly silicate, were very light and took months to settle down;
- the dust clouds were up to 8–10 km high;
- while the storm raged, surface temperatures fell by about 25°C, while the atmosphere instead absorbed solar radiation and warmed.

An infrared radiometer in the 8–40 micron range profiled surface temperatures in strips on succeeding orbits. A series of instruments was turned on by Mars 3 as it made close approaches over the surface, marking a track over the Martian landscape, the main sessions being on 15th and 27th December 1971, 9th January 1972 and 16th and 28th February 1972. The radio telescope could measure surface temperature down to 50 cm. The thermal map compiled by Mars 2 and 3 found that:

- temperatures ranged from −13°C to −93°C in the southern hemisphere, where summer was ending;
- temperatures fell to −110°C at the north pole in the midst of its winter;
- soil temperatures were unchanged during the day, but the surface cooled very quickly after sunset, suggesting a low level of conductivity and dry sandy soil;
- the Martian dark areas (or seas) cooled more slowly than the light areas (or continents);
- there were thermal hotspots where temperatures were much higher (up to 10°C higher) than surrounding areas;
- temperatures 0.5 m below the surface never rose above −40°C;
- soil temperatures in the equatorial belts averaged −40°C, but by 60°S latitude they were down to −70°C, irrespective of day or night.

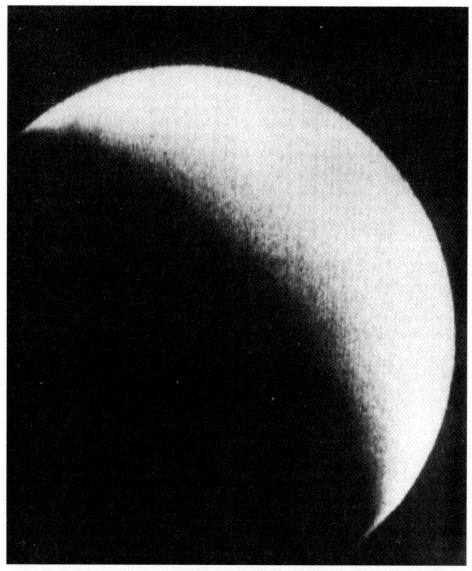

Mars 3 view of Mars

As for the surface, it was possible to estimate soil density at between $1.2\,g/cm^2$ and $1.6\,g/cm^3$. But, there were variations. In the *Cerberus* region, one of the warmer regions, density was calculated at $2.4\,g/cm^3$ and there were some places where concentrations were found of $3.5\,g/cm^3$. The surface was believed to be covered with dust made mainly of silicon oxide, but that the average depth of dust was only 1 mm.

The success of Mars 2 and 3 was eclipsed in the public mind in the West by that of Mariner 9. Although Mariner 9 was a small spacecraft compared with the Russian ones – it weighed only 1,031 kg, including fuel – it carried a full suite of scientific instruments and these matched Mars 2 and 3 findings in respect of the atmosphere, temperatures, pressures and environment around the planet. Mariner 9 mapped 70% of the planet, taking 7,000 images, providing a level of detail much deeper than the bleak images relayed during the Mariner 4, 6 and 7 flybys. Mariner 9 had scientists gasping as it sent back spellbinding pictures of craters, volcanoes, river beds, canyons and sand dunes, with more than a hint that the planet had a watery past. Mars had suddenly become an interesting place to explore. Presenting a paper at the International Astronautical Federation in Baku, Azerbaijan, the following autumn (1973), Vasili Moroz agreed that, although dry now, Mars could have supported large reserves of water – as recently as 25,000 years ago.

The scientific results from Mars 2 and 3 were published in papers over the following years, some focusing on the surface results, some on the atmosphere, others on the space environment around Mars. In summary, Mars 2 and 3 had accomplished much to the expansion of our knowledge of Mars, as this summary indicates:

Science from Mars 2/3: the main discoveries
Mascons affecting orbital paths up to 150 km.
Flattened poles, by 35 km.
Dust storms 10 km high.
Shock wave resisting the solar wind.
Surface temperature, pressure, density.
Composition of lower and upper atmosphere.
Carbon dioxide decomposes at 100 km.
Profile of surface temperatures by day and night.

Science from Mars 2/3: Mars surface
Temperature	$-110°C$ to $+13°C$
Pressure	5.5–6 mb, 1/200th that of Earth
Surface density	Between $1.2 \, g/cm^3$ and $1.6 \, g/cm^3$, in places up to $3.5 \, g/cm^3$
Composition	Dust mainly of silicon oxide, depth 1 mm

Science from Mars 2/3: atmosphere
Composition	Carbon dioxide, 90%; nitrogen, 0.027%; oxygen, 0.02%
Water vapour	10–20 µm, 1/2,000th to 1/5,000th that of Earth
Argon 40	0.016%

Science from Mars 2/3: environment
Base of the ionosphere: 80–110 km.
Very weak magnetic field.
Oxygen 800 km out, hydrogen up to 10,000 km.
Hydrogen corona between 10,000 km and 20,000 km out.

THE GREAT MARS FLEET: THE PROBLEM OF 2T-212

The Soviet Union decided to make an all-out assault on Mars during the next launch opportunity, 1973. The foremost consideration was the plan by the United State to follow the success of Mariner 9 with two large, expensive orbiter-landers in 1975, called Viking. The 1973 window represented a last chance for the Soviet Union to get back lengthy surface transmissions before the arrival of these American Vikings. The chances of success were rated highly, granted how close Mars 3 had come.

The Americans cannot have been under many misapprehensions as to Soviet intentions. 1973–5 was a period of détente between East and West and agreement had been reached between President Nixon and Soviet leader Leonid Brezhnev that there should be a joint manned mission in 1975 (Apollo–Soyuz). Joint working groups were set up to promote space cooperation in a number of areas, such as biology, medicine and planetary exploration. In one of their first agreements, the Soviet Union made available to the United States its data from Mars 2, 3 and Venera 8. In return, the Soviet Union asked NASA for its models of the atmosphere of Mars, predictions for conditions on Mars in spring 1974 and Mariner 9 maps of two landing zones.

The Mars window for 1973 was much less benign than that of 1971 and it would not be possible to launch combined orbiters and landers. Accordingly, the four spacecraft available were divided into two groups of two: Mars 4 and 5 were orbiters only (Mars 73S, S for *sputnik*), which – in addition to the scientific programme –were designed to act as relays for the landers, Mars 6 and 7 (Mars 73P, p for *pasadka*), whose mother ships would fly past the planet and not attempt to enter orbit [19]. An important change was that the landers were designed to transmit during the parachute descent, rather like the Venera probes as they entered the Venusian atmosphere. The spaceship was 4.2 m high, 2 m in base diameter (5.9 m wide with solar panels extended), a mass of 3,895 kg – of this, the mother spacecraft weighed 3,260 kg and the descent module 635 kg. Telemetry was transmitted on 928.4 MHz.

Experiments on Mars 4 and 5 (Mars 73S)
Magnetometer.
Infrared radiometer to study surface temperature.
Plasma ion traps.
Radio occultation device to profile density of atmosphere.
Radio telescope polarimeter to probe below surface.
Two polarimeters to characterize surface texture.
Spectrometer to study upper-atmosphere emissions.
Narrow-angle electrostatic plasma sensor to study solar wind.
Lyman-α photometer to search for hydrogen in the upper atmosphere.
Three cameras: Vega camera (52 mm), Zulfar (350 mm), panoramic.
Carbon dioxide photometers.
Water vapour photometer to detect water in atmosphere.
Ultraviolet photometer to measure ozone.
Stereo 2, to study solar emissions (France).
Zhemo: solar protons and electrons (France).
Photometers (four).
Polarimeters.

Experiments on Mars 6, 7 flyby modules
Telephotometer.
Lyman-α sensor to detect hydrogen in upper atmosphere.
Magnetometer.
Ion trap.
Narrow-angle electrostatic plasma sensor to study solar wind.
Solar cosmic ray sensor.
Charged particle detector.
Micrometeorite sensor.
Solar radiometer (France).
Radio occultation device.

Experiments on Mars 6, 7 landers
Cameras and telephotometer.
Thermometers.
Pressure, density, wind sensors.
Accelerometer.
Atmospheric density meter.
Mass spectrometer for atmospheric composition.
Activation analysis experiment for soil.
Mechanical properties soil sensor.

The Mars 4 and 5 orbiters carried the Vega, Zufar and panoramic cameras, altogether a big improvement compared with the 1971 missions. More film was carried than ever before, no less than 20 m in length, able to hold 480 pictures. Pictures could be taken at either 1/50 a second or 1/150 a second. Instead of just slow- and fast-scanning methods, no fewer than ten scanning rates could be chosen. In reality, three were used: preview (220×235 pixels), normal (880×940 pixels) and high resolution ($1,760 \times 1,880$ pixels). Pictures were relayed back at 6 kbps. The two panoramic cameras were of the optical–mechanical linear type, using a mirror to assemble an image that would cross $30°$ of the planet from one horizon to another, scanning at 4 lines/sec. These were an advance on the optical–mechanical cycloramic camera used on the landers, for the linear type was designed to be used from moving, flyby or orbiting spacecraft. Designed by Arnold Selivanov, they were first tested in lunar orbit by Luna 19 in 1971–2 and then applied to the 1973 Mars probes. It would take 90 min to assemble a panorama. The lander carried Vadim Istomin's spectrometer first flown on Mars 3.

Following the success of the Stereo experiment, more French equipment was installed. Stereo was carried again, along with a new experiment: built collaboratively this time between Audouin Dollfus of Meudon Observatory and Leonid Ksinformaliti, this was a visual polarimeter called the VPM-73 (Visual Polarimeter Mars 1973).

The preparations for the launch went relatively smoothly until spacecraft integration tests began. The electrical systems began to fail in all the four spacecraft under test and the culprit was found to be a humble transistor made in Voronezh called the 2T-212. To save money, the Voronezh plant had replaced the transistor's durable,

Arnold Selivanov

specified, gold leads with cheaper aluminium ones. The problem was that corrosion frequently caused aluminium leads to fail after about 1.5 to two years. The Voronezh transistors had been made a year earlier, so they were calculated as 'likely to fail' just as the probes approached Mars. All the spacecraft were full of 2T-212 transistors, so they would have to be almost completely disassembled, while a new, gold-plated production run was organized, which would take six months, long past the launch date. There was not enough time to disassemble the spacecraft and install a new run of quality standard circuits, so the alternative was a two-year delay.

An extensive risk assessment exercise was undertaken, examining failure rates in such transistors used in a number of other unrelated industries. The chances of failure were rated at 50/50. In the rush to beat the Americans, this was considered acceptable, so the Ministry for General Machine Building took the decision to go ahead regardless. On the part of the scientists, there may have been a fear that if the mission were postponed two years, it might never fly at all, in which case there could definitely be no chance of success [20].

In a further pre-launch crisis, the power control system of the third spacecraft (which would become Mars 6) short-circuited. It turned out that, despite supervision, a technician had managed to connect a hundred terminals the wrong way round. The conservative approach would have been to take out and replace the damaged power control system. Fearing that this would prevent the timely launch of the probe, the technicians reconnected the terminals the right away, checked that there was no apparent damage arising from the previous test and concluded the integration tests with no further incident.

ON THEIR WAY

All four spacecraft were launched successfully. Mars 4, launched 21st July, was 1.3m km away when Mars 5 left on the 25th. First course corrections were carried out on the 30th July and 3rd August, respectively. Four tracking ships were following the mission in the Atlantic. Mars 6 and 7 left Earth on 5th and 9th August, respectively. Radio Moscow proudly gave the position of the fleet's distance from Earth on the evening of the 9th: Mars 4 at 6.4m km, Mars 5 at 5m km, Mars 6 at 1.5m km and Mars 7 at 102,000 km away from Earth. It was the first time four out of four launchings had succeeded. Mars 6 made a course correction on 13th August, Mars 7 on the 16th August.

The feared computer problem duly materialized. Mars 6, the spacecraft which not only had bad transistors like the rest but had also suffered during the integration test, was first to go and ceased transmission at the end of September. This was actually announced at the time by Roald Sagdeev, although nobody paid much attention [21]. Ground control hoped against hope that this might be an isolated failure and, not convinced that the spacecraft had failed completely, continued to send up commands in the forlorn hope that the receivers might still be functioning. Then, Mars 7 suffered an early breakdown and only one transmitter operated. Next to go down was Mars 4, where two of the three channels of the computer failed, making a second mid-course correction impossible.

The full nature of the transistor problem was not revealed until *glasnost* 15 years later. There had been a big fanfare when the fleet departed Earth, but subsequent coverage could only be described as subdued and one had the sense that media expectations were deliberately lowered (nowadays this would be called 'news management'). But soon, in contrast to the Mars 2–3 missions, the progress of the Mars fleet was reported sparingly by the Soviet media, as if to dampen expectations. Now we know why.

As a result of its computer failure and the lack of a second course correction, Mars 4 could not arrive at Mars at the precise point required for entry into Mars orbit. It passed Mars at between 1,844 km and 2,200 km, just too far out for an entry burn. Despite this, ground controllers decided to take advantage of the close approach and carry out a photographic scan for the 6 min of closest approach. Mars 4 did take twelve Vega pictures and two panoramas of a swathe of the southern hemisphere, from west to east, including the intended Mars 6 landing area, sent back radio occultation data and continued to transmit from solar orbit, where it still orbits between 1 and 1.63 AU, 2.2°, every 567 days.

In a similar disappointment on 9th March, the Mars 7 computer gave a correct setting and command for the separation, firing and descent of the landing module and they appeared to be properly programmed [22]. Separation took place on schedule and all seemed to be going well, but the descent module's faulty computer then reversed the

Mars 4 passes Mars

command for ignition 15 min later. As a result, the lander sailed over the Martian surface at 1,300 km, following its mother ship into solar orbit. Mars 7 was almost certainly to have landed in crater *Galle* in the *Argyre Planitia* at 43°S, 42°W – for this was the other region where the Soviet Union had asked for Mariner 9 landing zone maps [23]. Occultation data were transmitted during the flyby. The French Stereo equipment worked from launch until May 1974. Mars 7 continued to transmit until at least September 1974, the last of the four to fall silent.

Arrivals at Mars, 1974

10 Feb	Mars 4
12 Feb	Mars 5
9 Mar	Mars 7
12 Mar	Mars 6

MARS 6 REACHES *MARE ERYTHRAEUM*

Ground controllers must have almost given up on the Mars 6 spacecraft, due to arrive three days later on 12th March: nothing had been heard from Mars 6 since the previous September. Although they did not know it, Mars 6 had, despite its transmitter failure, continued to respond to ground commands and to operate autonomously. The computer had orientated the spacecraft correctly over Mars and adjusted the trajectory correctly in the way intended by Mars 2 and 3. The lander was duly separated at 55,000 km/hr and the mother ship flew past the planet at 1,600 km and cruised on its solar orbit. Here, Mars 6 did carry out a radio occultation experiment whose results matched those of Mars 4 and 5 and these found a night side ionosphere with an electron density of 4,600 elements per cm^3.

The additional channel installed on the lander during the descent now proved its worth. The first ground control knew that the spacecraft was functioning was when a Doppler signal started streaming in from Mars 6 some 4,800 km out as the spaceship dived toward Mars at an angle of 11.7°. The landing spot was at 23°54'S, 19°25'W longitude, in the *Mare Erythraeum*.

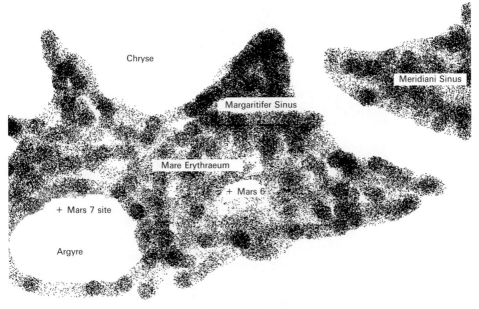

Mars 6, 7 landing sites

Separation of the 1,210-kg landing module had taken place at 5.7 km/sec some 45,000 km out. Entry into the Mars atmosphere took place at 11:53.38 on 12th March. Aerobraking slowed the descent module to 600 m/sec, and at 20 km altitude, prompted by the high-altitude radar, the parachute was reefed and then fully opened above the 645 kg lander. Now the scientific instruments came full on, relayed information to the mother craft at 256 kbps on frequencies of 122.8 MHz and 138.6 MHz. They came on for 149 sec, but were lost as the spaceship hit the surface at 61 m/sec, its standard speed of descent, but much higher than the intended touchdown speed of 6.5 m/sec. The signal fizzled out over a period of 1.3 sec.

It was 15:30 local time during the local spring. A light easterly breeze was blowing, 8–12 m/sec. We do not know precisely what went wrong. It is possible that it did not survive the impact, but these were toughly constructed machines and this speed was within tested limits. The period of greatest stress on the spacecraft was the very rapid descent from 75 km to 29 km, which took only a minute during blackout, so it had passed the most difficult phase. There could also have been a malfunction with the final landing rocket firing, though the low-altitude radar appears to have been in order. We simply don't know and Mars 6 was not the last lander to disappear on touchdown.

The following is a timetable of the landing:

11:39.07	Receipt of Doppler signal from lander, 4,800 km
12:06.20	Loss of signal due to blackout, 75 km
12:07.20	Acquisition of signal after blackout, 29 km, start of data

Mars 6 coming in to land

12:08.35	Parachute deployment
12:08.44	Parachute filled
12:11.04	Loss of signal at surface

At the time, the Russians announced that information had been transmitted during
the descent, but said nothing more. Claims were made that – as it descended – Mars 6
made the first direct measurements of the chemical composition of the atmosphere,
determining it to be mainly carbon dioxide but also finding argon. What appears to
have happened is that a few limited spectrometer data were returned during the
descent, the intention being to transmit the whole set after landing, so the figures
were based on the limited descent data. This issue was eventually resolved in the course
of a series of post-mission analyses. Study of the data went on for several years and
was revised many times over.

Live transmissions from the descent did broadcast data on pressure, temperature and atmospheric composition and as a result it was possible to compile a profile of the atmosphere [24]. The pressure readings show the pressure climbing from 2 mb high up in the atmosphere to 3 mb at the time of parachute deployment to 5.45 mb at the time of landing. Temperature readings show a rise in temperature from between $-131°C$ and $-109°C$ at 29 km to between $-63°C$ and $-58°C$ at 12 km and $-27°C$ at touchdown. Wind measurements were taken from 7.3 km altitude down to 200 m and they fluctuated between 12 m/sec and 15 m/sec. Deceleration measurements were taken, marking the G forces at entry, parachute deployment and landing, indicating forces of up to 9 G. It was calculated that the landing point was 3,388 km from the centre of the planet. Unlike 1971, visibility during the descent was good, so a photometer on board took colour filter images of the atmosphere during the descent [25]. Photometers had been carried on planetary probes since Zond 1 was sent to Venus in 1964 and the first such results had been returned from there by Venera 6 in 1969.

There were contradictory reports on the chemical composition of the atmosphere during the descent, determining it to be mainly carbon dioxide but also finding argon [26]. The chemical composition data fluctuated wildly, indicating high rates of water content and argon content, as high as 25% to 45% in places. The real rate is now known to be about 1.6%, so it is probable that the instrumentation was faulty. Sagdeev's view was the data were affected by the transistor failure, which made them erroneous [27].

Near-surface results from Mars 6

Wind	8–12 m/sec
Surface temperature	$-27°C$ (246°K)
Surface pressure	5.45 mb

Between Mars 3 and Mars 6, the USSR accumulated both descent data (Mars 6) and imputed surface data (Mars 2–3) and had laid a solid foundation for future landings on Mars. Yet, to this day Mars 6 remains the last Russian probe to reach the surface.

THE SHORT SUCCESS OF MARS 5

The other, albeit short, success of the great Mars fleet was Mars 5, which on 12th February 1974 entered Mars orbit of $5,154 \times 35,980$ km, period 24 hr 52 min 30 sec, inclination $35°19'17''$ precisely. No sooner had it arrived in Mars orbit than its housekeeping instruments reported a slow leak out of the pressurized, sealed instrument compartment. As a result, its planned three-month scientific programme was abruptly compressed into three weeks.

The cameras took about twelve pictures during each close approach, the main sweeps being on 17th February, 21st February, 23rd February, 25th February and 26th February. The five panoramic pictures scanned a 30° swathe with up to 512 pixels per scanline, taking panoramas covering swathes from 5°N to 20°S, 130°W to 330°W, which included the Mariner Valley. Of the 108 photographs taken, 43 were of

Mars 5

reasonable quality and five panoramas were assembled. They showed volcanoes, dried-up river beds, tectonic faults, sandy-bottomed craters and erosion of the land-scape. Soviet scientists carried out an examination of the features of Mars photo-graphed by Mars 4 and 5, finding what were interpreted as former river valleys, glaciated regions and rivers, with the inference that in the past there were many rivers on Mars and that the climate was once much warmer. Mars 5 was the first planetary probe to carry a photopolarimeter, designed to measure haze in the atmosphere and dust and sand grains on the surface. This French–Russian instrument was able to read details of the surface. For example, *Mare Erythraeum* was like a dusty lunar *mare*, while other regions had what seemed to be dunes (*Claritas Fossae, Thaumasia Fossae* and *Ogygis Rupes*). Within the craters Lampland and Bond, there were large boulders, but the wind had blown the dust off them.

Mars 5 made important discoveries about the atmosphere of Mars. Its instru-ments found five-fold fluctuations in atmospheric humidity along the flight trajectory. The Mars 5 photometer found water vapour up to 80 μm, compared with the 10 to 20 μm for Mars 3 (the higher rate was later confirmed by Viking, which found a rate as high as 100 μm over the polar regions). If condensed into a surface layer, this would have covered the planet's surface to a depth of 100 μm. Moreover, Mars 5's instru-ments found the levels of water quite variable from one region to another – by a factor of four. Mars 5 found an ozone layer 30 km above the equator and reconfirmed the earlier finding of atomic hydrogen 20,000 km out, the concentration being 1/1,000 that of Earth. The identification of argon was confirmed. A relatively high water vapour content (100 μm) was found in *Tharsis*. Sand storms were detected and followed [28]. Six carbon dioxide profiles of the atmosphere were done by the CO_2 photometer. Cirrus-like clouds were identified in the high atmosphere, as were yellowy clouds with fine dust particles. During perigee passes, surface temperatures were measured, from

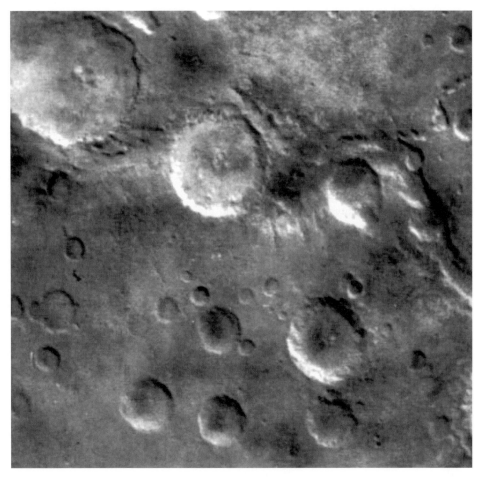

Craters from Mars 5

−1°C daytime to −73°C by night. Some surface pressures – presumably at lower locations than Mars 6's landing site – reached 6.7 mb.

Mars 5 carried a 256-channel gamma ray spectrometer which made 9 hr long data sets: 61m, 85m, 93m km out from Earth and six sets from Mars orbit. The gamma ray spectrometer switched on during seven low passes over *Thaumasia, Argyre, Coprates, Lacus Phoenicis, Sinus Sheba, Tharsis* and *Aria Mons* volcano, a big swathe of the southern hemisphere from 20°N to 50°S from *Amazonis* across to *Mare Erythraeum*, an area of 400,000 km^2. The uranium, thorium and potassium content of the surface corresponded to basic igneous rocks on Earth, but proportions varied, depending on the ages of the rocks concerned and whether they were old highland formations or younger volcanic formations [29]. The size of the rock analyzed ranged from grains of from 0.04 mm in wind-blown areas to 0.5 mm elsewhere. This found the following content of Martian rock:

Composition of Mars rock (Mars 5)

O	44%
Si	17%
Al + Fe	19%
K	0.3%
U	0.6 ppm
Thorium	2.1 ppm

Source: Surkov (1997) [29]

The issue of a magnetic field continued to attract attention. Mars 2 and 3's data indicated a magnetic field 0.015% that of Earth, but the Russian scientists were unsure whether this was a real, indigenous magnetic field or particles assembling in the shadow of the planet away from the sun and giving the appearance of a magnetic field. Now they seemed more certain that there was a magnetic field, located 15° off centre, currently in the southern hemisphere. Mars 5's data led scientists to draw up a map of three plasma zones around the planet. The solar wind hit Mars at mach 7 some 350 km out and made a bow shock.

Discoveries of Mars 5

Composition of Mars rock	
Variations in atmospheric humidity	
Levels of water vapour	
Ozone layer at 30 km	
Existence of argon	
Surface temperatures	−44°C to −2°C day, −73°C night
Exospheric temperature	21°C to 81°C, falling by 10° at 87 km to 200 km
Small magnetic field	0.003%
Earth atmospheric pressure	6.7 mb (combined with Mars 4, 6)
Electron density of ionosphere	4,600/cm^3 at 110 km

Mars 5 eventually failed after 22 orbits on 28th February when it depressurized. This may be less a problem than it suggests, for Russian spaceships were designed to collect substantial information intensively over a short period, rather than the longer term approach of the Americans.

Although the Mars fleet of 1973 was reported as largely a failure in the West, there were some important outcomes. Mars 6 provided an atmospheric profile, while Mars 5 added substantially to the results obtained by the Mars 2 and 3 orbiters. Impressive, scientifically authoritative and beautifully presented maps were made of Mars using the Mars 4–5 photographs: A.V. Sidorenko compiled, in *Poverknost Marsa*, a lengthy geographic, geological, geomorphological analysis of the scientific outcomes of the missions [30]. The specific outcomes of the Mars 5 mission were published in special issues of *Kosmicheski Issledovania* (vol. 13, #1, 1975; vol. 15, #2, 1977) while the outcomes of Mars 4, 5 and 6 were published in book form four years later as *Physics of the planet Mars*, by Vasili Moroz (Nauka, Moscow, 1978). Most went completely unnoticed in the West. Outside scientific circles, the Russian popular press did little to

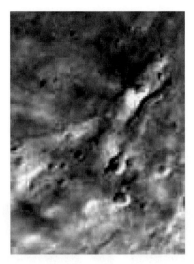

View from Mars 5

promote the outcomes of the Mars fleet, possibly because they fell short of their ambitions, but more likely for fear of drawing attention to the problem affecting the computer circuits.

MARS FLEET AFTERMATH

On the Soviet side, despite the scientific gains, the great Mars fleet was a great disappointment. The mood in the Soviet interplanetary programme was quite depressed in the summer of 1974. General Machine Building Minister Sergei Afanasayev decided that no more Mars probes of this generation would be built for Mars, although plans were already well under way to adapt the type for the nearer target of Venus. Afanasayev asked Sergei Kryukov to begin work, for introduction in the 1980s, of a new, universal generation of Mars–Venus explorers, which later became the UMV series (Universal Mars Venus). He also announced that the age of the transistor was over and that electronic systems should be used on the UMV probes thereafter.

The months that followed were a period of trauma in the Soviet space programme. In May 1974, chief designer Vasili Mishin, successor to Sergei Korolev, was dismissed and replaced by his great rival, Valentin Glushko. The new chief designer's first decision was to order a complete review of the centrepiece of the Soviet space programme, the moon programme. Korolev's N-1 rocket, dating to 1956, which had now been launched four times and came close to success on the fourth occasion, was cancelled, depriving the Soviet Union not only of its moon rocket but the ability to mount a Mars expedition and launch large space stations. Instead, Glushko focused his undoubted talents on a powerful new rocket, the Energiya and a space shuttle – the *Buran* – able to rival that of the United States. Construction of these

Valentin Gluskho

new projects was under way by 1976. The Lavochkin design bureau was not directly affected by these changes, except that its prefix was renamed NPO Lavochkin (NPO = Science and Production Organization).

The organizational upheaval of the time was reflected in the Institute of Space Research, IKI, although there there was a change in personnel rather than organizational architecture. With the approval of the government, Mstislav Keldysh had brought in a second director to IKI, Roald Sagdeev to take the place of Georgi Petrov. Born in Moscow in 1933, this mathematician turned nuclear physicist had spent his professional career in the science city – Akademgorodok – set up by Khrushchev in Siberia. The city followed in the tsarist tradition of welcoming exiled dissidents to Siberia: Khrushchev wanted them to have scientific (and some political) freedom away from the restrained environment of Moscow, but he (and they) knew that they were far enough away to be ignored. Forty-year-old Sagdeev brought with him from Siberia his open, questioning attitudes and soon began to shake things up. He brought in young scientists based on merit, not political affiliation; he recruited Jews at a time when some were politically suspect; fought for the modernization of computers and argued the merits of cooperation with the West, including the United States. He learned English, pressed the case for information exchange with foreign scientists and railed against the petty restrictions that made these things difficult, saving his strongest words for the government attitude to photocopiers, which had to be registered in case they were used improperly to undermine the government. In effect, he brought *glasnost* (openness) and *perestroika* (transformation, reform) before they were so labelled by Mikhail Gorbachev. His combination of charm, shrewd political judgement and patience enabled him to push back the barriers of what was possible [31].

Roald Sagdeev

The arrival of Glushko and Sagdeev provided an opportunity for a reconsideration of interplanetary missions. When Sagdeev arrived, preparation of the Mars fleet was well under way. The period after the Mars fleet was subsequently called 'the war of the worlds' [32]. Sensing a diminution in the political appetite for prestigious, competitive missions, scientists in the Academy of Sciences presented what they considered to be a more rational agenda. The Soviet Union, they argued, should concentrate on Venus, where it had built up a recognized expertise and elsewhere devise missions that *complemented* but did not *challenge* directly the United States. In arguing for a concentration on areas of existing expertise, Sagdeev found himself labelled a 'Venusian'. Not everyone agreed with the 'Venusians', pointing out that there were limits to what could be learned about Venus and that Mars was an intrinsically more interesting planet. Their camp was called 'the Martians'. Possibly to their surprise, the arguments of the 'Venusians' prevailed and the political leadership decided not to launch probes to Mars during the next launch window in 1975. Venus would be the focus of Soviet interplanetary exploration for the next ten years. The 'Venusians' won round one, though – as we shall see – the 'Martians' bided their time and prepared their next move.

The decision not to compete head to head with the United States in 1975 proved to be a wise decision, for little that they could have done that year would have bettered the astonishing achievements of the two American Viking probes. The Viking programme represented a significant step up in the scale of American planetary exploration, for these were large, complex spacecraft almost four tonnes in weight, close to the scale of the Soviet Mars probes. The programme was an expensive one, costing over $1bn and had roots in plans sketched far back in the early 1960s (the *Voyager* programme). Viking 1 and 2 mother craft entered Mars orbit, where their cameras could scan potential landing sites *before* the landing, a level of sophistication that the Mars 2–7 series could not match. This feature was wise, for the original landing sites turned out to be unsuitable and new ones had to be found. Viking's landing method of cone, parachute and rocket was similar to the Russian Mars programme, except that soft-landing braking rockets were used for the last full 2,000 m of the descent. Viking 1 landed on 20th July 1976, the seventh anniversary of the moon landing and Viking 2 a month later. They were lucky too, for had either landed just a few metres away from where they did, they could have hit boulders and overturned. They took colour pictures of the surface, used a mechanical arm to grab the soil and put it into a chemical laboratory and acted as weather stations on the planet. The mother ships carried out extensive mapping of the planet, building on the work begun by Mariner 9, and between them more than doubled the level of knowledge of the planet.

Following the success of the Lunokhod roving vehicles on the moon, VNII Transmash drew up a series of designs for roving vehicles on Mars [33]. These ranged from six-wheel vehicles to ski-walking machines to a machine with a trailer. Although they could well have taken the shine off the American achievements, the problem was to get these relatively large vehicles safely onto the surface in the first place, and the chances of successfully doing so were not rated highly.

The disappointing outcome of the great Mars fleet, the war of the worlds and the decision not to compete directly with the United States had broader implications. During the 1970s, leadership of planetary exploration passed decisively to the United States. Until then, the Soviet Union had always been first to send probes to planets. In 1972, the Americans had sent the first probe to Jupiter (Pioneer 10, 1972), Mercury (1973) and Saturn (Pioneer 11, 1973). The year 1977 would provide an alignment of the planets in such a way that it was possible to send a spacecraft swinging past Jupiter and on to Saturn, Uranus and Neptune on a mission popularly called 'the grand tour'. The United States prepared two spacecraft for this: Voyager 1 and 2 and their outstanding missions became history. In Korolev's time, the Soviet Union always had to be first. It was a reflection on the changed circumstances of the two rivals that such missions were not now even contemplated in the Soviet Union. The first design study of a Jupiter probe was not even undertaken until 1990.

But the best days were yet to come, for the victory of the 'Venusians' paved the way for what became the high summer of Soviet planetary exploration.

REFERENCES

[1] Grahn, Sven: *A Soviet Venus probe fails – and I stumble across it, http://www.svengrahn.ppe.se*

[2] Kerzhanovich, Viktor and Pikhadze, Konstantin: *Soviet Veneras and Mars – first entry probes trajectory reconstruction science.* Paper presented to the international workshop on planetary probe atmospheric entry and descent trajectory analysis and science, Lisbon, Portugal, 6th–9th October 2003.

[3] Objects from the Cosmos 482 launch. *Spaceflight*, vol. 44, September 2002.

[4] Venus – 470°C in the Sun! *Soviet Weekly*, 16th September 1972.

[5] Varfolomeyev, Timothy: The Soviet Venus programme. *Spaceflight*, vol. 35, #2, February 1993.

[6] Clark, P.S.: *Block D*. Paper presented to the British Interplanetary Society, 5th June 1999.

[7] Huntress, W.T., Moroz, V.I. and Shevalev, I.L.: Lunar and robotic exploration missions in the 20th century. *Space Science Review*, vol. 107, 2003.

[8] Perminov, V.G.: *The difficult road to Mars – a brief history of Mars exploration in the Soviet Union*. Monographs in Aerospace History, no. 15. NASA, Washington DC, 1999.

[9] Huntress, W.T., Moroz, V.I. and Shevalev, I.L.: Lunar and robotic exploration missions in the 20th century. *Space Science Review*, vol. 107, 2003.

[10] Ulivi, Paolo: *Exploration of the solar system*. Springer/Praxis, Chichester, UK, 2007, forthcoming.

[11] Tyulin, Georgi: Memoirs, in John Rhea (ed.): *Roads to space – an oral history of the Soviet space programme*. McGraw-Hill, London, 1995.

[12] Siddiqi, Assif: *The challenge to Apollo*. NASA, Washington DC, 2000.

[13] Much of our fresh information on the final stages of the Mars 2 and 3 missions comes from [8]: Perminov, V.G.: *The difficult road to Mars – a brief history of Mars exploration in the Soviet Union*. Monographs in Aerospace History, no. 15. NASA, Washington DC, 1999. For an original account, read Belitsky, Boris: How the soft landing on Mars was accomplished. *Soviet Weekly*, 15th January 1972.

[14] Turnill, Reginald: *Observer's book of unmanned spaceflight*. Frederick Warne, London, 1974. Surkov, Yuri: *Exploration of terrestrial planets from spacecraft – instrumentation, investigation, interpretation*, 2nd edition. Wiley/Praxis, Chichester, UK, 1997.

[15] Clark, Phillip S.: The Soviet Mars programme. *Journal of the British Interplanetary Society*, vol. 39, #1, January 1986.

[16] Huntress, W.T., Moroz, V.I. and Shevalev, I.L.: Lunar and robotic exploration missions in the 20th century. *Space Science Review*, vol. 107, 2003.

[17] Ulivi, Paolo: *Exploration of the solar system*. Springer/Praxis, Chichester, UK, 2007, forthcoming.

[18] Kondratyev, K.Ya. and Bunakova, A.M.: *The meteorology of Mars*. Hydrometeorological Press, Leningrad, 1973, as translated by NASA, TT F 816.

[19] TsENKI: *The 3MP series of spacecraft*, *http://www.tsenki.com*, 2005.

[20] Sagdeev, Roald Z.: *The making of a Soviet scientist*. John Wiley & Sons, New York, 1994.

[21] Mars probe misses target. *Flight International*, 26th February 1974.

[22] TsENKI: *The 3MP series of spacecraft*, *http://www.tsenki.com*, 2005.

[23] Klaes, Larry: Soviet planetary exploration. *Spaceflight*, vol. 32, #8, August 1990.

[24] Sagdeev, Roald Z.: The principal phases of space research in the USSR, in USSR Academy of Sciences, History of the USSR, New Research, 5, *Yuri Gagarin – to mark the 25th anniversary of the first manned spaceflight*. Social Sciences Editorial Board, Moscow, 1986; Kerzhanovich, Viktor V.: Mars 6 – improved analysis of the descent module measurements. *Icarus*, vol. 30, 1977; Kerzhanovich, Viktor and Pikhadze, Konstantin: *Soviet Veneras and Mars – first entry probes trajectory reconstruction science*. Paper presented to the international workshop on planetary probe atmospheric entry and descent trajectory analysis and science, Lisbon, Portugal, 6th–9th October 2003.

[25] Mars 5 and 6 flights analyzed. *Flight International*, 4th April 1974.

[26] Lewis, Richard S.: *The illustrated history of space exploration – a comprehensive history of space discovery*. Salamander, London, 1983.

[27] Sagdeev, Roald Z.: *The making of a Soviet scientist*. John Wiley & Sons, New York, 1994.

[28] Dollfus, A., Ksanformaliti, L.V. and Moroz, V.I.: Simultaneous polarimetry of Mars from Mars 5 spacecraft and ground-based telescopes, in M.J. Rycroft (ed.): *COSPAR Space Research*, papers, vol. XVII, 1976.

[29] Surkov, Yuri: *Exploration of terrestrial planets from spacecraft – instrumentation, investigation, interpretation*, 2nd edition. Wiley/Praxis, Chichester, UK, 1997; See also: Results of geological and morphological analysis of the images of Mars 4–5. *Icarus*, vol. 26, 1975.

[30] Sidorenko, A.V. (ed.): *Poverkhnost Marsa*. Nauka, Moscow, 1980.

[31] Thompson, Dick: The wizard of IKI. *Time*, 5th October 1987.

[32] Sagdeev, Roald Z.: *The making of a Soviet scientist*. John Wiley & Sons, New York, 1994.

[33] Kemurdzhian, A.L., Gromov, V.V., Kazhakalo, I.F., Kozlov, G.V., Komissarov, V.I., Korepanov, G.N., Martinov, B.N., Malenkov, V.I., Mityskevich, K.V., Mishkinyuk, V.K. *et al.*: Soviet developments of planet rovers 1964–1990. CNES & Editions Cepadues: *Missions, technologies and design of planetary mobile vehicles*, 1993, proceedings of conference, Toulouse, September 1992.

6

The high summer of Soviet planetary exploration, 1975–1986

The cabin touched down on a gently rolling stony plateau. On impact, it raised a cloud of volcanic dust. The sky was orange, while the lava, stones and sand were greenish yellow.

– Soviet account of the landing of Venera 13

NEW GENERATION

Venera 8 marked the limit to what the Soviet Union could achieve with the 3MV series and the 8K78M launcher. The 3MV series had been kept going for Venus probes over 1965–72 even though it was no longer used for Mars probes. Although Minister Afanasayev had decided to move on to a new type of interplanetary spacecraft, it had already been decided to adapt the Mars 2–7 type of spacecraft to the exploration of Venus. This was formally called the 4V1 series, although in practice the term was little used.

The new generation of Venus spacecraft were likewise cylinders, with an engine at their base, lander on top and two large solar panels, essentially the same as the Mars 2, 3, 6 and 7 landers. Like the Mars 2–7 spacecraft – to which they were similar – the new Veneras were large: 2.8 m tall, with a solar panel span of 6.7 m. The probes had a series of pipes, used both to dump excess heat and to cool the landers before their arrival. The main engine, called the KTDU-425A, could be relit seven times for mid-course and orbital manoeuvres. The overall structure weighed in at 5,033 kg, including the entry probe of 1,560 kg and within it the lander of 660 kg (Venera 10 figures). The data transmission rate trebled. The mother craft carried on its frame:

- Two solar panels, with magnetometer, gas nozzles for orientation.
- Cold and hot radiators.

The 4V1 series (Venera 9)

- Gas bottles for the attitude control system.
- Wide-beam antenna, narrow-beam antenna.
- Sun reference sensors, Canopus[1] sensor and Earth sensors.

The mother craft would detach a lander as before, but it would come into the atmosphere at a much shallower angle, 18–21°, reducing the G loads from 400 to 500 G on the earlier spacecraft to a more tolerable 150–180 G.

The lander stood 2 m high and had a double hemisphere able to withstand 2,000°C and 300 tonnes of pressure. Before landing, it would be cooled to between

---

[1] Second brightest star in the sky – a supergiant only visible from the southern hemisphere on Earth.

−100° and −10°, so that its instruments could last longer when it was warmed up. The lander was radically re-designed. It looked like a mixture of a pressure cooker and a kettle with a metallic ring about it – this was a titanium aerobrake or disk brake designed to slow the spacecraft as it descended. Before, the parachute had been kept on until landing. Now, the parachute would be severed at 50 km: the disk brake would take its place in such a way as to get the cabin to the surface as fast as possible, but not so fast as not to survive touchdown. Over 1973–4, the new disk brake was tested in wind tunnels and in drops from helicopters and aircraft from altitudes as high as 14,000 m. The spacecraft model was also dropped onto simulated Venusian soil made from foam concrete.

The base of the lander was a shock-absorbing, crushable ring, on which were placed scientific instruments. The main part of the lander was supported by struts from the ring. Two downward-facing cameras with goldfish bowl lenses were located just under the disk brake, one on each side. They were angled in such a way that each would take a single 180° panorama, two completing 360°. The cameras were pinpoint photometers of the type used on the lunar soft-landers (Luna 9, 13, etc.) and on the Mars 2, 3, 6 and 7 landers. Location of the camera system was a compromise. If it were placed on top, like the lunar and Mars landers, most of the surface would be obscured by the airbrake. Placed below the airbrake, it would be difficult for it to see far into the distance. So it was placed below the airbrake, in such a way as to look down not only at the surface detail, but capture the horizon to the side, though not straight ahead. A special pressure window had to be built to protect the cameras from the extreme pressures and temperatures. Because data transmission rates were slow, 256 bits/sec, the 512×128-pixel images would be scanned intentionally slowly at one line every 3.5 sec, but as long as the lander lasted half an hour a full panorama would be sent.

A helical antenna was wound around the upper cylinder. Two pipes led from the lander through the disk brake to the mother craft, and these were used to pump a blast of cold air into the lander just before the descent. The orbiters carried a series of experiments, including optical–mechanical linear cameras of the type flown on Mars 4 and 5 to make a photographic study of the planet.

An extensive suite of instruments was developed for the new Veneras. For the first time, nephelometers – designed by Mikhail Marov – were installed to measure the nature of the cloud layers during the descent through the atmosphere.

Instruments on Venera 9–10 landers
Panoramic telephotometer
Photometer to measure chemical composition of atmosphere.
Instrument to measure radiation in atmosphere from 63 to 18 km.
Temperature and pressure sensors.
Accelerometer to measure G forces during descent.
Anemomometer to measure wind.
Gamma ray spectrometer to measure radioactive elements in rocks.
Radiation densitometer.
Mass spectrometer.

The new Venera lander

Instruments on Venera 9–10 orbiters
Panoramic camera.
Infrared spectrometer.
Infrared radiometer to measure cloud temperatures.
Photopolarimeter.
Spectrometer.
Magnetometer.
Plasma electrostatic spectrometer.
Trap for charged particles.
Ultraviolet imaging spectrometer (France).

VENERA 9 LANDS ON A MOUNTAINSIDE

Venera 9 and 10 left for Venus on 8th and 14th June 1975, beginning an unbroken run of successful launchings to the planet. Venera 9 made two course adjustments, a

Venera lander in shell

velocity change of 12 m/sec on 16th June and one of 13.5 m/sec on 15th October. Venera 10 also made two, on 21st June and 18th October (14.5 m/sec and 9.7 m/sec, respectively).

Venera 9 approached Venus on 20th October. The entry probe was released. The main spacecraft flew on, fired its engine for a change of 247.3 m/sec to avoid a Venus impact and then made a larger burn of 922.7 m/sec to become the first spacecraft to orbit the planet Venus, from $1,500 \times 111,700$ km at $34°10'$. The final orbit was $1,510 \times 112,200$ km, 48 hr 18 min, inclination $34.17°$ (though it is uncertain if the final figures came from more accurate measurements or actual changes by the motor).

The lander plunged into the atmosphere at 10.7 km/sec. The 2.4 m aluminium heatshield was released at 64 km, when a 4.4 m metallic parachute opened from three drogue chutes, which reduced the speed to 250 m/sec. The lander was then cooled to $-10°$C for the final descent. The entry angle was a shallow $20.5°$, intended to reduce the stresses of descent.

At 50 km – in the clouds – the parachute was jettisoned and then the disk brake began to function as the main decelerator. Instruments relayed a stream of data to the orbiter for onward transmission to Earth, some of the instruments being interrogated twice a second. After a further 75 min the cabin was on the surface, the touchdown assisted by shock absorbers, landing at 8:28 a.m. on 20th October. Final impact speed was 7 m/sec or 20 km/hr. A gentle breeze was blowing, between 1.4 km/hr and 2.5 km/hr. A sudden drop in light level at touchdown suggests the lander kicked up a cloud of dust. The landing point was latitude 31.7°, longitude 291° in *Beta Regio*. By this time, the Venera 9 orbiter was on its first orbit and had swung into position over the landing site, ready to relay messages back to Earth.

Surface experiments were initiated immediately. Based on the original probe used by Luna 13 to test the lunar surface in 1966, a density meter jabbed into the soil. The main experiment was photography. After 2 min, caps dropped off the protective cameras of the spacecraft. Venera 9 carried a 10,000-lux floodlight, to be set off if the light meter determined it to be necessary, for many of the scientists had expected cloud, dust and darkness. The Venera 8 photometer had indicated a gloomy surface, but this was not the case here (the sun was 54° above the horizon for Venera 9, much higher than its predecessor). The 5.8 kg, 5 W camera was 90 cm above the surface, half way up the 2 m tall lander and took images of 517 lines. The picture took 30 min to transmit, through high-pressure windows made of quartz 1 cm thick and the objective was to send one full panorama. Images were interspersed with other scientific data.

The picture was received on Earth an hour later. It surprised the scientists, who had expected to see only rocks in the immediate foreground. Instead, the horizon could be seen 300 m away and the immediate area was full of round and curved rocks on a dark surface. Venera 9 came down in a young mountainscape 2,500 m above the datum or reference line, the equivalent of sea level (datum is defined as a radius of 6,051 km of Venus) on the side of a hill or volcano. It could have been on a hill, with a slope of 15–20°, or possibly the side of a volcanic crater, scattered with sharp and rounded stones and slabs from a recent eruption. There was no dust and rocks could be identified up to 100 m away. It was possible to make out sharp stones about 35 cm across with the soil between them. Some sunshine was reaching the surface, for some rocks had shadows. The rocks were not eroded by wind, so either they were young, or there was little wind erosion on Venus. The radioactive chemical composition of the rocks was 0.3% potassium, 0.0002% thorium and 0.0001% uranium, with a rock density of between 2.7 and 2.9 g/cm^3, more like basalt than the granite indicated by Venera 8.

Temperatures inside the craft began to rise and – when they reached 60° inside after 53 min – transmissions ceased (the orbiter was also moving out of contact at this point). The only disappointment was that one of the cameras failed, so only one 180° panorama was returned, covering one side of the spacecraft only.

Venera 10

Venera 10 touched down on 25th October. Even with a shallow entry of 23°, temperatures still reached 12,000°C and G forces 168 G. The parachutes were dropped off

Venera 9, first panorama

at 49 km, the nephelometer recording an end to clouds at 30 km. At 42 km, the temperature was 158°C, pressure 3.3 atmospheres, rising to 363°C and 37 atmospheres 15 km above the surface, as it fell under its disk brake. Meantime, the main spacecraft also braked into Venusian orbit.

Venera 9 first image

Venera 9 rocky terrain

Venera 10 landed on a much less exotic landscape. Landing point was latitude 16°, longitude 291°, a similar longitude but different latitude, also in *Beta Regio* some 1,400 km away. The instruments reported a surface pressure of 92 atmospheres and a temperature of 465°C.

It was at once apparent that Venera 10 had come down on an older landscape – a rolling plain with outcrops of hard, decrystallized magmatic rocks having undergone considerable chemical weathering [1]. Some of the rocks were called 'pancake rocks' because of their shape. The penetrometer measured the density of the soil, finding it to be 2.7 g/cm^3, reflecting its rockiness. The lander was leaning backward on a 3 m slab, probably giving a more distant horizon for the single panorama. Again, one camera failed, so there was a view from only one side. Surface transmission times were 65 min for Venera 10, ending only when the mother spacecraft passed out of reception range.

Meantime, the mother ship made a deflection burn of 242.2 m/sec, followed by a braking burn of 976.5 m/sec to enter an orbit 1,400 × 114,000 km, leading to a final orbit of 1,620 × 113,900 km, 49 hr 23 min, inclination 29.5°. Its first apoaxis brought it over the Venera 10 landing site, ready to transmit its signals on to Earth.

Venera 10 first panorama

Venera 10, first image

The two orbiters went on to make observations from Venus orbit in a programme that lasted until 22nd March 1976. In-orbit experiments covered the planet's atmosphere, clouds, composition, temperature, particles and magnetic fields. An ionosphere was detected. Using cameras developed by Arnold Selivanov, the orbiters took panoramic pictures of Venus, each image taking in 1,200 km swathes. Most important were those panoramas taken using filters, for they could make out the cloud structures and even penetrate clouds. By 5th November, Venera 9 had sent back seven and Venera 10 five such pictures, sufficient to discern the fact that Venus had flat areas and mountains. They were relayed back to Earth at both slow and fast mode. Venera 9 took a total of 17 such image sets by the end of the year, making out detail as small as 6.5 km on the planet. Only a few of these high-quality images appear to have been published. Venera 9 and 10 carried a 32 cm radio radar which mapped 55 strips of the surface in swathes 200 km wide by 1,200 km long and a resolution of up to 20 m.

As for the other experiments, Venera 9 found that Venus, although a non-magnetic planet, nevertheless had a magnetic plasma tail. The planet's ionosphere formed a magnetic barrier preventing the solar wind from entering the planet's atmosphere and causing an impact wave on the sun's side and a tail on the other [2]. The results of the orbiting mission were published in *Pravda* the following 21st February. They found upper clouds to have a temperature around −35°C and that nighttime clouds were about 10°C higher than daytime. Cloud temperatures declined as one descended, but there were many variations at 55–66 km. Venera 9 was the first spacecraft to suggest that Venus had three distinct layers of cloud.

Venera 10, second image

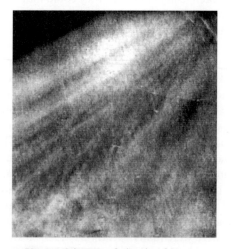

Venera 9 image of clouds of Venus

Science from Venera 9, 10 orbiters

Cloud base	30–35 km, layered to 64 km
Cloud layers glow at night	
High levels of corrosion due to sulphuric acid, especially upper atmosphere	
Bromide and iodine vapour in lower clouds	
Clouds denser at the equator, spiral toward the poles	
Cloud temperatures	35°C light side, 45°C dark side.
Highest cloud temperatures	40–50 km above surface.
Temperature before cloud entry	−35°C
Carbon dioxide to water ratio at 38 km	1,000:1

Science from Venera 9 lander

Sufficient natural light (10,000 lux) on the surface to take pictures without floodlights	
Dust stirred up on landing	
Wind	0.4–0.7 m/sec (less than 10 km/hr)
Temperature	480°C
Soil density	2.7 g to 2.9 kg/cm^3
Pressure	90 atmospheres
Cloud layers	3

Science from Venera 10 lander

Wind	0.8–1.3 m/sec.
Temperature	465°C
Pressure	92 atmospheres
Surface density	2.8 g/cm^3

Venera 9 and 10 were outstanding successes. Two probes had been launched and two had fully accomplished their missions. The use of the new design for the Venus probe had been vindicated and the lander design had worked perfectly, including the new disk brake. Both probes had transmitted for lengthy periods from the surface, providing not only important information but the first pictures from the surface. Additional data had been obtained from the orbital missions – another first. They were the last probes witnessed by one man who had a long association with the Soviet interplanetary programme and who had calculated the original trajectories to Venus. Mstislav Keldysh had retired in 1975, dying three years later of a heart attack in his country home. A quiet man who wielded his considerable power lightly, his unusual combination of political and mathematical skills did more than any other Soviet scientist to make the early missions to the planet possible. An oceanographic research ship was named after him and later a research centre.

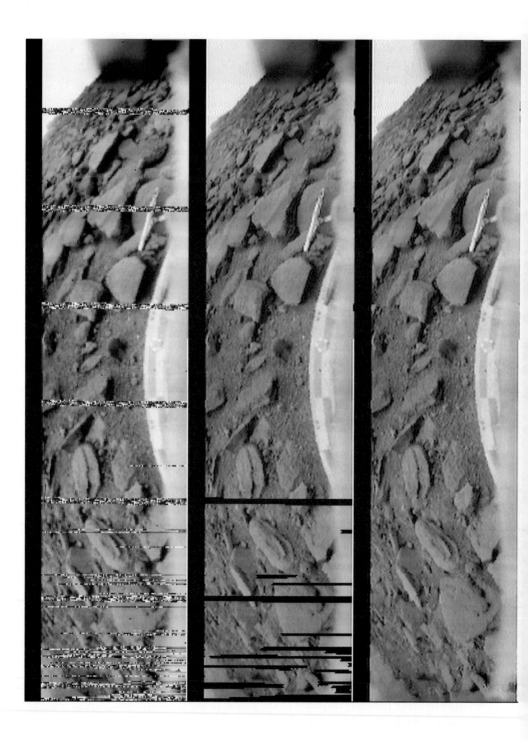

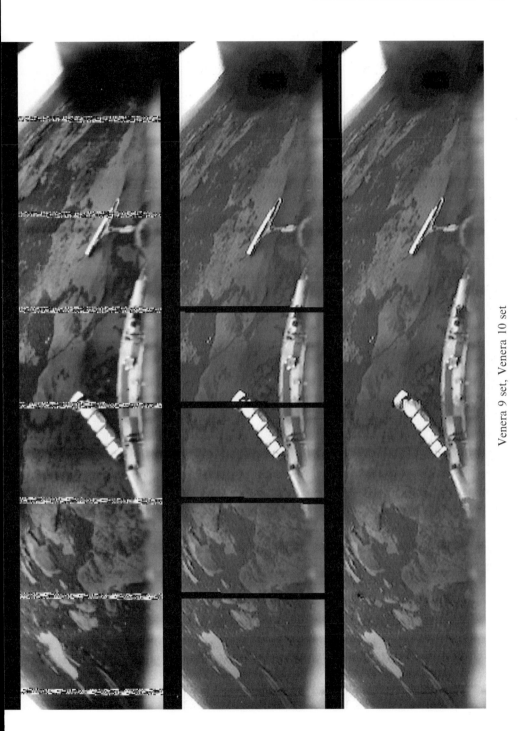

Venera 9 set, Venera 10 set

VENERA 11, 12: THE FIRST SOUNDS OF ANOTHER WORLD

Repeat missions were flown by Venera 11 and 12 on 9th and 14th September 1978, still called the 4V1 series. Course corrections were carried out on 16th and 21st September, respectively. The launch announcement was low key, saying little about the purpose of the mission except that the two craft were identical but carried equipment different from their predecessors. French scientific equipment was carried on Venera 11. Scientific director of the double mission was Vladimir Kurt, who contributed the gamma burst meter.

Venera 11 and 12 were the first to benefit from significant upgrades in the Soviet tracking system. Originally, this comprised the 1960 deep space control system of 64 m dishes in Yevpatoria and Ussuriisk, supplemented by the 32 m Saturn dishes in the mid-1960s. Now, new 70 m dishes called the P 2500 system were installed in Yevpatoria and Ussuriisk, as well as two 64 m dishes in Moscow in Kaliazin and Bear's Lake. Overall, this was called the Kvant D system.

The probes were launched very late in the launch window and at a time when conditions were becoming unfavourable. This reduced the amount of fuel that could be carried, which meant that the mother craft would pass Venus rather than enter orbit, dropping landers on a high-velocity flypast. Although no orbital mission was

Venera 12 in preparation

Venera 12 at Baikonour

possible, the mother ships would actually remain longer in the line of sight of the lander than an orbiter, improving the length of time for the receipt of surface transmissions. Their weight was given as 3,940 kg, but 4,715 kg when fuelled.

On the way to Venus, instruments were used to locate the source of and characterize gamma ray bursts. One of them – called Konus, developed by Vladimir Kurt – picked up 27 gamma ray bursts and aimed to answer the question: Where were these bursts coming from? Accordingly, their sources were triangulated between the two Veneras and a solar observatory then in Earth orbit called Prognoz. Konus also identified 120 solar flares and 20 X-ray eruptions from the sun. High-energy particles were measured by the KV-77 instrument. During one session of two hours, Venera 11 and 12's instruments measured charged particles coming from the sun in a powerful burst of solar flares. Other instruments were used to measure the solar wind's composition and its interaction with interplanetary space.

Venera 11 and 12: in-flight instruments
Omni-directional gamma ray and X-ray detector (France).
Konus cosmic ray detector.
KV-77, for high-energy particles.
Plasma spectrometer.
Ultraviolet spectrometer.
Magnetometer.
Solar wind detector.

Venera 11, 12: lander instruments
Panoramic colour camera.
Gas chromatograph.
Mass spectrometer.
Gamma ray spectrometer.
Lightning detector.
Temperature and pressure sensors.
Anemometer.
Nephelometer.
Optical spectrophotometer.
X-ray flourescent spectrometer.
Accelerometer.
Soil penetrator (PrOP-V).

Venera 12 made a final course correction on 14th December and Venera 11 on 17th December. Arriving in reverse order, the Venera 12 lander entered the atmosphere on 21st December at 11.2 km/sec, plunged through the clouds and dropped its parachute at 49 km. As the airbrake came into play and as the atmosphere thickened, the descent rate slowed from 50 m/sec to 8 m/sec on landing. The impact churned up dust which took 25 sec to settle and blew away under the 1 m/sec surface winds. Venera 12 transmitted for 50 min during the descent and a further 110 min from the surface. The descent of Venera 12 was the longest of any of the series, providing ample opportunity for measurements to be made.

Venera 11, arriving 25th December, was dropped off to land on the dark side of the planet, making a dustless touchdown and transmitted for 95 min. The landers came down 800 km apart. The landing point for Venera 11 was latitude 14°S, longitude 299° and for Venera 12, the landing point was latitude 7°S, longitude 304° – both east of *Navka Planitia*. Transmissions ended in both cases only when the mother craft went out of range.

A strange thing happened to both probes at 12.5 km altitude above the surface. Although the engineers called it 'an anomaly', all the instrument readings went off scale and there was a sudden electrical discharge from the spacecraft.

In both cases, the mother ships flew past at 35,000 km, making deflection manoeuvres, but no braking manoeuvres. The Venera 12 mother ship was the first Soviet spacecraft to observe a comet. After passing Venus, in 1980 it took – with its ultraviolet spectrometer built by Soviet and French specialist – images of comet Bradfield. The Konus gamma ray detector instrument, combined with the French instrument Signe, tried to ascertain the nature of high-energy bursts from Pisces, thought to be some form of pulsar, first detected by the Soviet Earth-orbiting observatory Prognoz 7 in March 1979: 150 gamma bursts were recorded.

For many years, Venera 11 and 12 were the source of another of the mysteries of Soviet planetary exploration. No pictures were published, although transmissions from the surface were definitely received. At one stage, it was stated clearly that tele-panoramas were returned, even that they showed a flat rocky landscape, but they were never seen. It was then explained that pictures could not be taken since both landed by night, but this explanation was implausible, for the previous Veneras

Venera 12 lander

carried floodlights and presumably these could have been used. At one stage it was suggested that the whole surface package had failed and that scientific data came entirely during the descent.

In reality, Venera 12 and 11 represented an ambitious step forward, carrying colour cameras, a laboratory to drill and analyze the soil and other new experiments. The new 70 m receiving station at Yevpatoria was specifically brought into operation for this mission, because of the much higher transmission rate expected from the cameras, up from 256 bits/sec (Venera 9–10) to the 3,000 bits/sec expected from Venera 12 and 11. The new system would enable colour pictures to be transmitted fast, one every 14 min, compared with one black-and-white image every 30 min. Venera 11 and 12 carried the first systems designed to drill and analyze the rocks of Venus on board. This assignment went to the General Construction Design Bureau of Vladimir Barmin (1909–1993). Barmin was a close colleague of Sergei Korolev and a member of the original Council of Designers of 1946. He was the constructor of the cosmodromes, a task of enormous proportions involving the heaviest Earth-moving and digging machinery in the world. His bureau's skills at this level were matched by his ability to make precision, small-scale machinery. He had built the drilling rig used to obtain a core sample from the Sea of Crises on the moon with Luna 24 in 1976.

Ground control waited expectantly – but there were no pictures. Improved seals had been installed on the lens caps on Venera 12, but they were sealed so well that they would not eject. When Venera 12 drilled the soil, the sample was not correctly transferred to the examination chamber, possibly due to a leak in the system. The PrOP-V penetrators on arms, designed to test the strength of the surface, were

Vladimir Barmin

wrecked by the roughness of the landing. PrOP-V stood for *Pribori Otchenki Pro-khodimosti Venus* (literally, instrument for evaluating the characteristics of the surface of Venus). But it was not a mobile rover, like the PrOP-M developed for the Mars 3 and 6 landers.

When Venera 11 landed several days later, ground control sat helpless when exactly the same thing happened again. Despite this, there were scientific results from Venera 11 and 12. Because of the lack of surface pictures, Venera 11 and 12 were largely written off in the West. This probably did less than justice to the scientific return from the missions. For their part, the Russians devoted an entire issue of *Kosmicheskiye Issledovaniya* (Cosmic Research) to the outcomes of Venera 11 and 12 (the September/October 1979 issue).

Each transmitted for over an hour and a half from the surface, a remarkable achievement in itself. Basic surface temperature and pressure data were returned. There were some minor differences in the descent package, for Venera 11 carried a nephelometer while in its place Venera 12 carried an aerosol analyzer.

Most of the data came during the descent and these elaborated what had been gathered by Venera 7 to 10. The 10 kg gas chromatograph, a new instrument, took in nine samples during the descent between 42 km and touchdown, drawing them into porous materials where their elements were measured. This new instrument had been developed by an expert in planetary formation and atmospheres, Lev Mukhin (b. 1933) of the Institute for Space Research. This found some quantities of argon in the atmosphere. Detection of argon was important, because argon is a decay product of potassium which in turn is an indirect indicator of vulcanism.

The X-ray fluorescent emission spectrometer measured cloud particles during the descent from 64 km to 49 km when it was destroyed by the temperature. It found sulphur ($0.1 \, \text{mg/m}^3$), chlorine ($0.43 \, \text{mg/m}^3$) and iron ($0.21 \, \text{mg/m}^3$). Venera 12 made

Lev Mukhin

probably the most precise measurement of Venus' atmosphere. The main element was of course carbon dioxide, 97%, with nitrogen 2%, but Venera 12 was able to measure the minor elements as well.

During the descent, Vladimir Istomin's mass spectrometer was turned on – once the probes were clear of the cloud layer at 25 km altitude. This mass spectrometer was of the same type as that developed for the Mars 3 and 6 landers. Venera 11's found 0.5% water vapour at 44 km, falling to 0.1% at 24 km, while Venera 12's found none at all. Twenty-two gas samples were taken, a chemical breakdown of each one being relayed as they descended, giving detailed readouts of the proportions of nitrogen, argon, neon, krypton and xenon gases. Every sample was measured several times, after which the container was pumped out to a vacuum before the next one was taken in.

The nephelometer transmitted details of the clouds during the descent. The final cloud layer was between 48 km and 51 km, with mist below. The photometer measured the amount of sunlight reaching the surface. The spectrophotometer found that the clouds were relatively transparent: although 3–6% of sunlight reached ground level, the sun was so dispersed you could never see it directly from the surface.

Science from Venera 11, 12

Surface pressure	88 atmospheres (Venera 11) and 80 atmospheres (Venera 12)
Surface temperature	446°C (Venera 11) and 500°C (Venera 12)
Presence of argon-36–40 and argon-56 in the atmosphere	1/200 times smaller than Earth
Large-scale, intense thunder and lightning	

Venus atmosphere: the Venera 12 results
Major elements
Carbon dioxide   97%
Nitrogen   2–3%

Minor elements
Water   700–5,000 millionths
Argon   110 millionths
Neon   12 millionths
Krypton   0.3–0.8 millionths
Oxygen   18 millionths
Sulphur dioxide   130 millionths
Carbon monoxide   28 millionths

Perhaps the most intriguing experiment was Groza, a Russian word meaning 'thunder'. As far back as the early 1960s the director of Kharkhov University Observatory, N. Barbashov, predicted that Venus would have thunderstorms a thousand times more powerful than anything on Earth [3]. The idea was taken up by Leonid Ksanformaliti (b. 1932), a 1956 graduate of Leningrad Technical University, subsequently at Abastumani Observatory. He had designed radiometers for Mars 3 to Mars 7 and the spectrometers that analyzed the clouds of Venus from the Venera 9 and 10 orbiters. The Groza instrument was carried to measure the sounds of Venus – wind, thunder and lightning – from an altitude of 62 km to the surface. It certainly found them, all the way down from 32 km to 2 km altitude. Venera 11 counted up to 25 lightning strikes a second and Venera 12 detected a total of 1,200 strikes altogether. After it landed, a massive thunderclap reverberated around the Venera 12 site for 15 min and probably affected the whole planet. One thunderstorm occupied an area of 150 km horizontally by 2 km vertically. Because of the high altitudes of the clouds, lightning strikes were likely to be cloud to cloud, rather than cloud to ground.

The Groza measurements were the first sound recorded on another world. In a missed public relations opportunity, the Russians never thought of releasing the groza sounds. Only when the European Huygens probe landed on Titan in 2005 and sent back a microphone recording of the sounds of Titan did it occur to anyone to see if it were possible to put the Venera 11/12 sounds on tape. The Planetary Society contacted Leonid Ksanformaliti and asked him to retrieve the old instrument recordings with a view to digitizing them and making them a sound that could be heard by Earthlings.

Venera 11 and 12 were matched by two outstandingly successful American probes, Pioneer Venus 1 and Pioneer Venus 2. Both were launched well ahead of the Veneras: Pioneer Venus 1 in May and Pioneer Venus 2 in August. They arrived ahead of the Veneras in early December. Pioneer Venus 1 entered orbit, the first American spacecraft to do so, circling at $145 \times 66,000$ km, 24 hr and sent back transmissions for no fewer than 14 years. Pioneer Venus 2 comprised a bus spacecraft with one large 317 kg and three small 91 kg probes. They became the first American spacecraft to reach its surface. The multiple spacecraft were designed as entry probes, rather like the actual missions carried out by Venera 4–6 and were not intended to soft-land. Although not designed to do so, one of the smaller probes reached the

Leonid Ksinformaliti

surface and transmitted from there for 67 min. The orbiter carried a radar altimeter which determined the basic parameters of the planet's surface to a resolution of 75 km and found two defined highland areas. Other instruments measured the planet's clouds and winds.

VENERA 13, 14: DRILLING THE ROCKS

Sorting out the difficulties arising from Venera 11 and 12 took some time, longer than the period before the next launch window. Venera 13 and 14 were modified to carry out the mission originally intended for Venera 11 and 12 and received the code 4V1M series. The new landers – each now weighed 760 kg – took advantage of Soviet advances in new heat-resistant technologies and lubricants able to work in extremes of heat.

Venera 13 was launched on 30th October 1981 and Venera 14 on 4th November 1981. First course corrections were made on 10th and 14th November respectively. Venera 13 made a second correction on 21st February. Due to an inaccurate first correction, Venera 14 was obliged to make two further corrections: on 23rd November and 25th February. *En route* to Venus, Venera 13 and 14 detected 20 gamma ray bursts and ten solar flares. As was the case with Venera 11 and 12, the mother craft would pass by the planet and not attempt to enter orbit. Venera 13 was aimed at hilly, ancient granite crust and Venera 14 at younger, lowland, lava flood plain, regional types which between them covered 80% of the planet. As was the case with Mars some years earlier, there was collaboration between American and Russian scientists in the planning of the landing sites. The Russians accepted recommendations from the US Geological Survey, moving the landing sites to sites which the Americans felt would be more interesting.

Venera 13–14 lander instruments
Gravity meter.
Nephelometer, to measure the concentration of aerosol particles.
Mass spectrometer, to determine the chemical composition of the atmosphere.
Gas chromatograph, to determine the chemical composition of the atmosphere.
Spectrometer and ultraviolet photometer to measure solar radiation in the atmosphere and water vapour content.
Telephotometer, to take pictures.
Penetrometer on the landing ring to test the strength of the surface.
X-ray fluorescent spectrometer to drill and test the chemical composition of the soil.
Lightning detector (Groza 2).
Pressure and temperature indicators.
Radio spectrometer to analyse electric and seismic activity.
Hydrometer/humidity sensor to detect water vapour content.
Experimental solar detector to measure light intensity.
Accelerometer.

Venera 13, 14: mother craft instruments
Gamma burst detector (France).
Cosmic ray detector.
Solar wind detector.
Magnetometer.

Venera 13 arrived on 1st March 1982, after travelling 130m km. The probe was released at a distance of 33,000 km out and entered the atmosphere at 11.2 km/sec. The first instruments started transmitting at 110 km and the rest of the descent package activated when the parachutes opened at 63 km. The parachute released the 760 kg craft at 47 km and the spacecraft disk-braked over the hilly, rolling land of *Phoebe Regio*, between 1,500 m and 2,000 m above the reference level.

Vsevolod Avduyevsky's instrument called BISON measured the precise speed of impact. Venera 13 stirred up a cloud of dust as it landed – in fact, it bounced from its 7.5 m/sec first impact on the crumbly rock and then settled back again. The bottom ring of the lander had razor-shaped teeth so as to make sure it would get a proper grip on the soil. The temperature outside was 457°C and the pressure 89 atmospheres, but temperature inside was only 30°C and pressure one atmosphere. Venera 13's landing point was latitude 13.2°S, longitude 310°, east of *Navka Planitia*. Initial wind speed was 0.3–1 m/sec.

Venera 13 photo sets

The camera lens fell off at once – properly this time – and work began. Eight panoramas were taken, using red, green, blue and clear filters in turn until such time as a colour image could be assembled. The camera comprised a mirror within a tube, looking down on the surface with a 37° field of view. The landing ring had green, white, blue, grey and black colours painted on it for reference. The transmission rate was much improved, permitting one line to be sent every 0.82 sec and the resolution was improved to $1,024 \times 252$ pixels. A fast scan was done first, in case the lander succumbed quickly, followed by a slow one. Venera 13 revealed a scene of stony desert, with an outcropping of bed rock and depressions in between with loose fine-grained soil [4].

The main experiment now began. First of all, the empty chemical laboratory on board was scanned, so as to set a baseline reading. A mechanical ladder straight away extended onto the surface and began to drill the rock using screw drills. The drills went 30 mm into the surface and extracted a $2\,cm^3$ sample. The samples were carried in and blown through three locked chambers of decreasing pressure and temperature (the last was at 30°C). There they were analyzed by an X-ray fluorescent device, irradiated by plutonium, uranium-235 and iron-55 and scanned 38 times. The system had to work against time, ever conscious that the probe would soon yield to pressure, temperature and sulphuric rain. The first drilling and scanning was done within 4 min and the overall retrieving and measuring of the sample took 32 min, the period of life for which the lander was guaranteed to operate. The pressure inside the chamber was 1/2,000th that of outside and had a temperature of 30°C.

Venera 14 landed on 5th March a 1,000 km away, coming down on a low-lying basaltic basin 500 m above sea level, with a pressure of 93 atmospheres, temperature 465°C. Venera 14's landing site was 13.2°S, 310° longitude and the lander transmitted for 57 min. Both probes experienced the same electrical anomaly as their predecessors coming through the 12.5 km mark. The pictures were different, looking more like icing on a baked cake surface or as Moscow Radio put it more scientifically, 'wrinkled

Venera 14 photo sets

brownish slate like sandstone'. It seemed to be a harsher, rockier, stonier, more weathered plain with fine-grained layered rock, no loose surface soil and a continuous rocky outcrop stretching toward the horizon. The drilling arm reached down, scooped up the rock into a hermetically sealed chamber and likewise put it through X-ray and fluorescent analysis, taking 20 spectra. The lander took in a 1 cm^3 soil sample drilled from a depth of 30 mm.

Each lander carried a seismometer: Venera 13 detected nothing, but Venera 14 noted two events, though they could have been the operation of the lander itself. Venera 13 and 14 carried the PrOP-V penetrometers originally developed for Venera 11 and 12. Venera 13's worked but Venera 14's had the misfortune of deploying right on top of the ejected camera lens cap and analysts had to disentangle the data of where the penetrometer ceased to measure the strength of the lens cap and that of the surface itself!

The rock analysis was soon back: 45% silica oxide, 4% potassium, 7% calcium oxide, with a general composition of basalt, basaltic rocks, alkaline potassic salts. Venera 13's rocks were alkaline basalt of the type found in the world's oceans and reckoned to be typical of two-thirds of Venus. They were not the granite that Venera 8 had led them to expect. Scientists believed that what they saw at the Venera 13 site was 'the planet's old crust – the surface in the area was badly eroded, except for outcrops of the bedrock and heavily strewn with crushed fine-grained material'.

Remarkable as these achievements were in drilling and analyzing the surface, they had none of the dramatic impact of the cameras. Eight images were taken by each probe, orange–brown in colour. Venera 13 sent back separate panoramas in red, green and blue showing a stony, rolling plateau and the curved horizon in the distance with stones, pebbles and flat rocks scattered all over it. There was close-grained soil studded with stones up to 5 cm in diameter. Above was an orange sky. Due to the effects of the atmosphere, which absorbed the blue part of the spectrum, the sky appeared to be orange while the lava, the stones and the sand were greenish

Venera 13: First image

yellow. A small clod of regolith was churned up during landing. Because the lander operated for 127 min, there was sufficient time for successive pictures to show wind being blown off the lander at a speed of 0.3–0.6 m/sec. Although the wind blew off the lander, Venusian winds are generally light and not enough to cause significant surface erosion.

The second Groza now came into play. The microphone was able to detect the lens caps falling off the cameras, the drilling sound (quite loud), the laboratory at work and in the background the wind, blowing lightly, at less than 50 cm/sec.

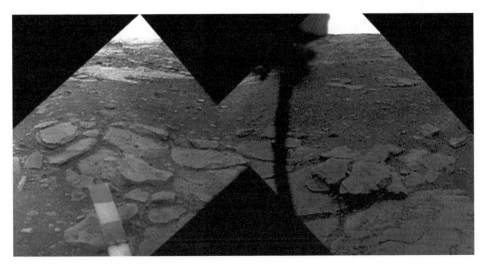

Venera 13: Second image

Venera 14: First image

VENERA 13 AND 14 OUTCOMES

Venera 13 and 14 were more than landers, for they were virtual surface laboratories. The extensive range of instrumentation enabled a broad range of scientific measurements to be made and for direct comparison between the two sites. Planetary geologists determined that the basalts of the two sites were quite different. Venera

Venera 14: Second image

13's site was up to 2 km above sea level, flat with weather-beaten lava flows with fine rubble. The cloud of dust raised on landing suggested that the ground was covered with wind-eroded volcanic lava.

By contrast, the Venera 14 landing site was freer of rubble, but the surface had plates in five separate layers without any marked elevations and a fairly straight horizon. The smooth, flat, rock-strewn plain suggested a long, continuous sedimentary process. It was smoother, stratified and less weathered, meaning that the structure was younger. The sulphur content was 0.3%, suggesting it was the youngest Venus rock yet found. The main chemical elements of the rocks were determined to be magnesium, aluminium, silica, potassium, calcium, manganese and iron. A table was compiled of the respective chemistry of the two landing sites:

	Venera 13 (%)	Venera 14 (%)
Magnesium	11.4	8.1
Aluminium	15.8	17.9
Silicon	45.1	48.7
Potassium	4	0.2
Calcium	7.1	10.3
Titanium	1.59	1.25
Iron	9.3	8.8

Within that, there were distinctions, for Venera 13 had 45% silica oxide, 4% potassium oxide, 7% calcium oxide compared with Venera 14's almost 49% silica oxide,

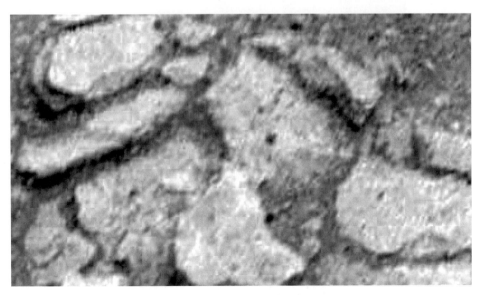

Venera 13, details of rocks

0.2% potassium oxide and 10% calcium oxide. The strength of the soil was calculated by the PrOP-V: Venera 13's corresponded to fine-grained sand.

Discoveries of Venera 13, 14
Chemical composition of the Venus rock.
Two minor seismic events.
Characterization of two landing sites.
The blowing away of dust by wind.
Strength of the soil.
Dehydration of the atmosphere.
Dispersal of water vapour in the atmosphere.

Meantime, as they passed by the planet, the mother ships measured the chemical composition of the atmosphere and found a new chemical compound called SF_6. They found the atmosphere was now highly dehydrated, with the suggestion that water had in the past played an important role in the formation of the cloud layer. The mother craft continued to transmit and on 11th May detected a burst of gamma ray radiation. They detected 89 cosmic gamma bursts and 300 solar flares. Venera 13 fired its engine on 10th June 1982, imitating a manoeuvre planned for the forthcoming VEGA mission. Venera 14 did so too, much later, on 14th November 1982.

Venera 13 ridge on horizon

The descent package added to the knowledge of the profile of the atmosphere collected by Venera 11 and 12. When Venera 13 and 14 first hit the uppermost fringes of the atmosphere at 90 km, there was a pressure of 0.0005 atmospheres and a temperature of $-100°C$. By the time they were down to 75 km above the surface, the atmosphere had grown to 0.15 atmospheres and the temperature had risen to $-51°C$. The probes solved the problem of discrepancies in the measurement of water vapour by previous probes. Venera 13–14 found that water vapour, small though it could be, was uneven, being most concentrated between 40 km and 60 km, lowest in the very high atmosphere and immediately above the surface. No fewer than 6,000 spectrographs were taken during the descent, detecting chlorine and sulphur. The water content was estimated at 0.2% at an altitude of 48 km. During the descent, the gas chromatograph made precise measurements of the level of water (700 ppm), oxygen (4 ppm), hydrogen (25 ppm), hydrogen sulphide (80 ppm) and carbonyl sulphide (40 ppm). Confirming previous descents but in more detail, the nephelometers identified at this stage three distinct layers of cloud:

- dense clouds at 57 km and above;
- a transparent mid-layer at 50–57 km;
- a final denser layer at 48–50km.

The amount of sunlight reaching the surface was 2.4% (Venera 13) and 3.5% (Venera 14). Venera 13 and 14 were the third set of successful double landings.

VENERA 15 AND 16: THE PLAN

Between them, Venera 7 to 14 had radically extended the knowledge of Venus's surface in eight distinct spots and profiled the atmosphere on the way down. Quite different spacecraft were prepared for the next set of missions, called the 4V2 series. Venera 15 and 16 left Earth on 2nd and 7th June 1983. When they were launched, it was announced that both would be orbiters and would make 'prolonged observations'. Later in the summer, it was revealed that they were carrying radars.

Venera 15 and 16 were each the same weight, 5,300 kg, but, instead of a lander, that entire part of the spacecraft had been converted into a radar system. A 300 kg radar panel was installed at the top, accompanied by transmission equipment and a large dish antenna on the side, 1 m wider in diameter than the previous probes. The central bus was lengthened by 1 m. The radar, called Polyus, was 6 m from end to end and 1.4 m wide. The designer was the man who found the signals from Venera 7, Oleg Rzhiga at the Institute of Radio Engineering and Electronics. The radar was built in such a way that it would paint the surface of Venus with microwaves for 3.9 msec every 0.3 sec, the images then being stored in two alternating computers. About 3,200 overlapping images would be taken on each pass, transmitted back to Earth and fed into a SPF-SM supercomputer built by Yuri Alexandrov at the Institute of Electronic Control Computers. The solar panels were twice the size of

Venera 15, 16 series

the previous Venera, in order to feed the electrically thirsty radar system. These were the changes:

- More fuel, so tanks were made 1 m longer.
- Solar battery power was doubled, with, in effect, double panels on each side.
- The parabolic antenna was enlarged by 1 m.
- Transmitter capacity was increased 30 times, with information flow of 108 kbytes/sec.
- Installation of a 6 m wide radar designed to image at 1–2 km resolution, measure height to 50 m and compile a radiothermal chart.
- Radiometric system Omega (25 kg).
- Radar altimeter with an accuracy of 50 m [5].

The transmission and receiving systems were developed by the radio electronic bureau of the Moscow Engineering Institute of Alexei Bogomolov (b. 1913). The fuel load was, instead of the normal 245 kg, a record 1,985 kg to insert the craft into Venus orbit and make subsequent manoeuvres. The high volume of information to be transmitted would stretch the full tracking system of Yevpatoria, Bear's Lake and Ussurisk to the limit. Even then, there were complications. The entry point of Venera 15 and 16 into Venus orbit, circling the planet 4° apart, had to be timed in such a way as to maximize line-of-sight transmissions from Venus to Soviet territory. Several course corrections

Alexei Bogomolov

would be required to adjust the path to achieve the best line-up with Yevpatoria. With Venus's slow rotation of only 1.48° every day, it would take the probes from November to July to carry out the mapping mission.

Venera 15, 16: instruments
Radar.
Omega radiometric system.
Fourier infrared spectrometer (GDR).
Radio occultation device.
Dispersion.
Cosmic ray detectors.
Solar wind detector.

In-flight instrumentation included instruments to measure cosmic rays and solar radiation. For the orbiting part of the mission, there was a 35 kg Fourier infrared radiometer made by a team of scientists in the Berlin Space Research Centre in the German Democratic Republic, under the leadership of Dieter Ortel, to remote-sense the atmosphere, cloud layers and thermal radiation and measure surface temperatures. About 60 measurement points would be made on every pass. There was a radio occultation experiment called Dispersion.

Venera 15, 16 radar system Polyus

VENERA 15, 16: THE MISSION

Venera 15 and 16 both altered course twice. Course changes were made for Venera 15 on 10th June and 1st October and by Venera 16 on 15th June and 5th October. They blasted into orbit around the planet on 10th and 14th October 1983, the third and fourth Soviet spacecraft to do so. Venera 15 had travelled 340m km, holding 69 communication sessions *en route*. Venus was 66m km distant from Earth at the point of arrival.

They corrected their orbits around the planet on 17th and 22nd October, respectively, to achieve their operational orbits (1,000 × 65,000km, 1,440 min, 87.5°, with perigee at 60°N) and later modified them in turn on 9th April 1984 and 21st June 1984. The operational mission started for both on 11th November 1983.

This was how they worked. On each northbound pass, the spacecraft turned their radars on, the intention being to map the northern part of the planet as the spacecraft flew over the pole in the course of 15–16 min imaging each orbit. Each radar image had a resolution of 2 km and covered 1m km^2. What they did was start to image at 80°N, go over the pole, down to 30–35°N, compiling 8,000-km long radar strips of 120 × 160 km width. At the same time a narrow-beam altimeter acquired height data

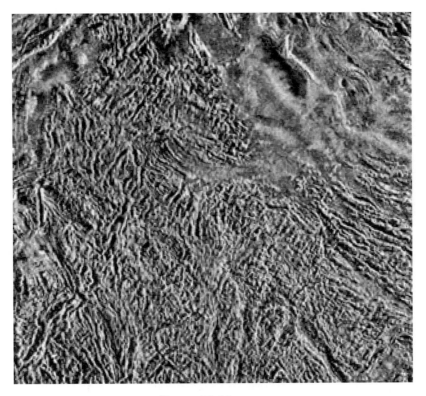

Venera 15, 16 tessera

to an accuracy of 50 m, and altogether 415,000 readings would be made. The orbiting spacecraft moved 155 km along the equator each revolution, or about 1.48° longitude each day, requiring eight months to map the planet's northern region.

The reason for this pattern was that the northern regions of Venus were geologically the most interesting, for they contained the two upland areas of the planet, *Maxwell Montes* and *Ishtar Terra*. A certain amount of low-resolution radar mapping of Venus had been done by Earthbound radars, but the polar regions could not be covered from Earth. The American Pioneer Venus mission in 1978 had made a basic topographic radar map of Venus, but the resolution was only between 75 m and 200 m. Venera's radars offered 50–100 times greater detail than what was possible from Earth. The total area to be mapped was 115m km². Venera 15 and 16's radars looked down at an angle of 10°.

Images were relayed to the Institute of Radiotechnology and Electronics (IRE) which had been carrying out Earth-based radar surveying of Venus since the mid-1960s. The images were then transmitted to Earth at a rate of 108,000 bytes/sec during 100 min daily communications sessions. Processing each image on the ground took eight hours and the amount of computer tape eventually built up to a length of 600 km! At one stage, Venus went out of range behind the sun and so transmissions

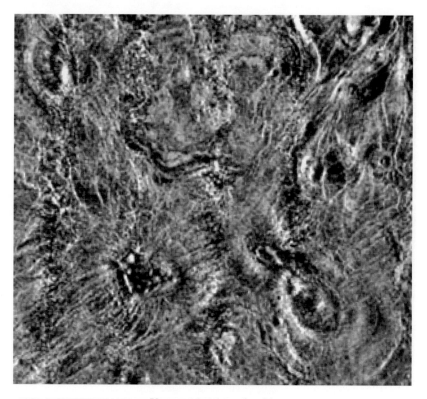

Venera 15, 16 arachnoids

could not be received. Venera 16 made an orbital plane change to return to re-map the missing region. The images from Venera 15 and 16 were interpreted by a 15-person team at the Vernadsky Institute for Geochemistry and Analytical Chemistry, led by Alexander Bazilevsky.

Although the radar images were the most important part of the downlink relays, they were not the only part, as this breakdown of communication sessions indicates: the transmissions could be divided into 116 sessions for the Polyus radar and Omega, 38 spectrometer and 90 radiophysical downlinks (Venera 15), and 174, 4 and 83, respectively, for Venera 16. More than 1,000 Fourier spectrograms were received. After radar mapping, the next most important objective was to compile a temperature profile of the atmosphere, determining how the planet conserved and distributed its heat and a thermal map was compiled.

Radar mapping concluded on 10th July 1984 and Venera 15 ran out of gas soon after. Venera 16 kept transmitting data from its other instruments for some time afterwards. Venera 15 had held 441 communication sessions (372 in Venus orbit) and Venera 16 had held 419 sessions. By this time, the maps covered the planet in a line from 24–33°N to the poles. Venera 15 and 16 mapped over 40% – that is, 120m km^2 – of the planet's surface. The probes extended their coverage southward in the remain-

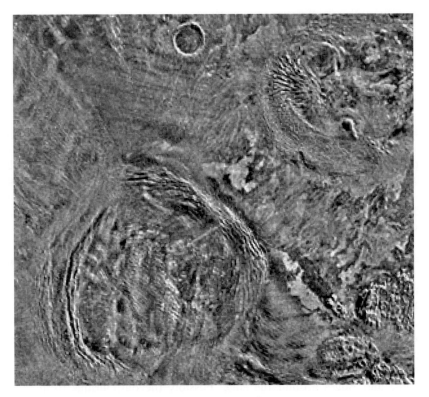

Venera 15, 16 coronae features

ing period of the mission. Communications with Venera 15 ended in March 1985 and Venera 16 not long afterwards.

VENERA 15, 16: OUTCOMES

The results of Venera 15 and 16 were published by the Soviet Academy of Sciences as a 27-mosaic radar, relief and geological atlas of the planet in 1987. The map comprised 115m km^2 of all latitudes north of $30°N$. The radar maps were originally compiled by hand and then later by computer.

Venera 15 and 16 found no fewer than 146 impact craters and a 400,000 km^2 lava floodplain, as well as plateaux, depressions, volcanic domes, shields, plains and valleys. Some features were adjudged to be asteroid craters up to three billion years old. Images showed craters, ancient ring structures, elliptical craters and volcanic domes. Image interpreters marked out a 'parquet' terrain, faults, impact craters, volcanic calderas, arachnoids (spidery features), depressions, furrows, linear ridges, mountains 12 km high, dome-shaped hills and scarps. In *Ishtar Terra*, the size of Australia, one mountain was found to be over 13 km high, higher than Everest, while in *Maxwell Montes* there was a circular crater nearby 96 km across. The northern pole was found to be relatively depressed, with most below the datum line. The most remarkable discoveries were those of large annular structures 400 km in diameter, called *coronae*, which had no direct comparators with Earthly or other planetary geology. Venera 15 and 16 data suggested that the surface of Venus was not unlike the volcanic plains of the moon and Mercury, primarily basaltic lava, with rock eroded by wind. The age of the surface was estimated at one billion years.

Most of the northern hemisphere comprised flat terrain, similar to the lunar one, the plains of Mars or the ocean floors of many parts of Earth. These plains were interspersed with occasional upland areas, like *Ishtar Terra*, rising 4,000 m high above the plains. On the plains, the radars detected what appeared to be outflows of basaltic lava which, even in Venus's high temperature, quickly formed a cooling crust as they vented. This was the broad picture, but some extraordinary detail was made out. The radars found:

- dome-shaped hills rising above the plains, sometimes a couple of kilometres across;
- ridges of 100–200 km in length, sometimes with parallel ridges between 8 km and 14 km alongside;
- craters, from 8 km diameter upward, the largest being Klenova, 144 km across. Venus's thick atmosphere was able to burn up small meteorites, but anything above 1 km across would get through;
- about 30 coronae (e.g., *Bachue, Anahait, Pomona*), with a jumbled relief in the middle and ridges around the edge, possibly the outpourings from volcanic hotspots – these could have been domes that bubbled up and subsequently collapsed;
- long linear ridges;
- large, volcanic, calderic circular depressions up to 280 km across (e.g., *Collette, Sacajawea*);
- a large depression at 70°N, subdivided into two regions of undulating plain (*Snegurochka Planitia*) and a flat and ridge belt plain (*Louhi Planitia*);
- linear ridges and grooves around *Laksmi Planum*;
- ridges with massifs of 8 km, 9 km, 12 km (*Vesta Rupes*);
- ridges and grooves like parquet flooring (*tessera* after the Greek for 'tile');
- haloes around volcanoes;
- elliptical structures measuring 300 × 500 km with ridges and grooves (*Tethus*);
- lava flows, seen as streaks on mountain slopes (*Theia* and *Rhea Montes*);
- troughs (*graben*) in upland areas (*Beta Regio*).

Finding craters was a breakthrough, for lunar scientists had used the density of craters to measure the age of the moon's surface (the more craters, the older it was). The Veneras were not able to detect active vulcanism, but crater density can be a good guideline to the respective effects of vulcanism and impacts. American lunar scientists developed various crater density models and from them Soviet planetologists calculated the age of Venus's surface as between 300 million and one billion years old,

Venera 15, 16 volcanic caldera

depending on the model adopted. They concluded that Venus's vulcanism was much less vigorous than Earth's but that it was still going on. In the case of some craters, it was not possible to tell the difference between volcanic or impact craters.

The other factor forming Venus's surface was tectonism, the way in which the crust pushes into other parts of the crust (on Earth there is mountain formation as one continent pushes against another). Tectonism appeared to have created two sets of uplands on Venus: ridges and grooves and smooth domes. The largest uplands were belts (e.g., *Maxwell Montes*, standing 4,000–10,000 m above their surrounding plains) and ridges and grooves, with ridges often intersecting at right angles, causing what was originally called parquet terrain but later tiled terrain or *tessera*. Smooth domes, such as *Beta Regio*, were sometimes 1,000–2,000 km across, standing 3–5 km over their surrounding plains, with gentle slopes and elevation. Venera radar images suggested that Venus experienced a prolonged period of basaltic vulcanism and intensive tectonism – a bit like Earth, but without the benefit of oceans, water or weathering. Venus showed an astonishing absence of erosion, probably due to the constant nature of its dense atmosphere [6]. Venus gave the impression of being more active than Mars, but less active than Earth.

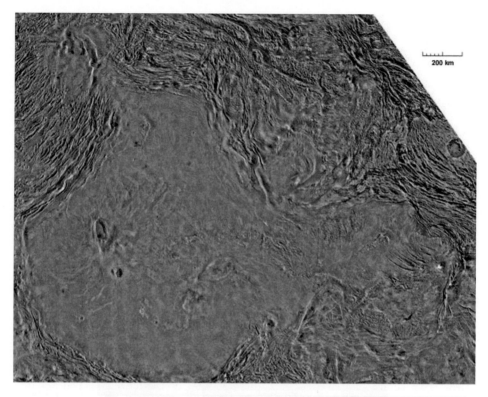

Venera 15, 16 *Maxwell Montes*

The Academy of Sciences map divided the landscape into flat plain, undulating plain, plain with mounds and ridges, parquet, ridge belts, ridges, hills, craters, valleys and ledges. Venus was characterized as:

High mountainous plateaux	8%
Smooth lowlands	27%
Rolling uplands	75%

Source: Surkov (1997) [4]

The rolling uplands are now thought to be the preserved old crust of Venus, covered by crater and reforming features, while the lowlands – because they have no impact craters – are thought to be young. Areas were considered upland if more than 500–2,000 m above the datum reference level (*Ishtar Terra* and *Maxwell Montes*).

The other instruments were able to draw up a temperature–altitude–pressure profile from between 60 km and 90 km in the cloud layer. The probes found thermal variations between latitudes: for example, it was warmer in a belt at 79–80°N than at 50°N. The average surface temperature was measured at 500°C, but the infrared spectrometer on Venera 15 (16's failed) found hotspots (localized thermal anomalies) with temperatures of 700°C above the surrounding areas, possibly volcanoes. Several localized thermal anomalies were also found. One was in *Beta Regio* from 281° to 288° longitude and from 17°N to 32°N; the second thermal anomaly was at 0° to 15°N latitude from 60° to 70°E, in *Maxwell Montes*; in both cases the areas were substantially cooler than the surrounding terrain.

As for the atmosphere, the Fourier spectrograms found polar clouds 5–8 km lower than equatorial clouds and that thermal radiation and temperatures were lower in the polar regions. The main observable strata of cloud were observed from orbit to lie from 47 km to 70 km, broadly matching the earlier descent data.

Venera 15, 16: achievements and discoveries
Mapping of 40% of surface (northern and polar regions).
Thermal characterization of atmosphere: identification of anomalous coldspots, hotspots.
Confirmation of level of main cloud layer.
Typology of planetary surface (mountain plateaux, lowlands, rolling uplands).
Typology of individual surface features.

The mapping of Venus was a technological triumph and scientific success. The members of the team responsible for Venera 15 and 16 were rightly awarded state prizes. The Venus maps were remarkable not just for the individual features discovered, but because they provided such a rich level of detail as to permit planetary geologists to make strides in developing a considered interpretation of Venus's past history and its present state of development and evolution. Later, contour maps were published. A heat atlas, arising from the GDR infrared spectrometer, was published.

The American spacecraft *Magellan* effectively completed the mapping of Venus. Launched by the space shuttle in May 1989, *Magellan* entered Venus near-polar orbit of 294 × 8,450 km, 3.2 hr in August 1990, eventually burning up in October 1994. *Magellan* had a 3.7-m dish able to provide a resolution of 120 m, enabling three-dimensional maps of the surface to be compiled. *Magellan*'s maps had a much higher resolution, a better viewing angle and covered the entire planet, not just the northern region.

The Soviet maps were poorly appreciated in the West in the 1980s and they were once belittled as providing little advance over American Earth-based observations. This was not NASA's view, its director describing the maps as first class and achieving a level of technological development which had not been fully appreciated [7]. Although the achievements of these Veneras were publicized [8], they were much less noticed than they deserved. A number of American scientists participated as guests of the Vernadsky Institute on the Venera 15 and 16 mission and joint symposia were organized after the mission, the two countries exchanging their Venera 15–16 and *Magellan* data.

Nitrogen supplies (see bottles) determined the length of the mission

TOWARD VEGA: THE BALLOON PROJECT

The zenith of the high summer of Soviet planetary exploration was the VEGA project. This was the last use of the Mars/Venus spacecraft introduced in 1969. For such a successful project, it had an unusual history of twists and turns.

For many years, the Soviet Union expressed interest in a mission to explore Venus by balloon. In the late 1970s, plans were discussed with French scientists to mark the 200th anniversary of the flight of the Montgolfier Brothers (1783) by dropping a large, red, commemorative balloon into Venus's atmosphere, with a 25 kg gondola carrying scientific instruments. There had been a long history of Soviet–French interplanetary cooperation and French equipment had flown on the 1971 Mars missions. In 1974, the Soviet Union opened discussions with the French on a balloon mission to Venus in 1981, but the French government would not support the project.

The idea of a balloon mission was resurrected at the 1979 annual Soviet–French cooperative meeting, held that September in Corsica. Two 10 m diameter aluminium balloons would be dropped at 50 km, fall for 10 km and circle at around 45 km altitude, the intention being that they should fly at the altitude where the clouds were thickest. It was an adventurous project and would take up to a tonne of payload

VEGA balloon concept

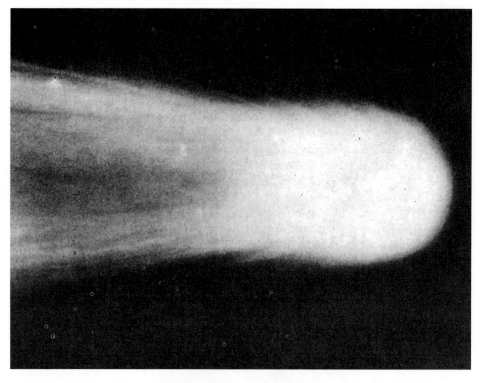

VEGA's target: comet Halley

weight and a gondola of 150 kg. It was tentatively called the Venera 84 mission. The mother craft would go into orbit around Venus, act as relay for the balloons and then go on a six-month mission observing the planet.

In the meantime, the world's scientists became increasingly conscious of the arrival in the inner solar system in 1985–6 of comet Halley, which was in a 76-year orbit around the sun. Japan announced plans to send two small spacecraft toward Halley while the European Space Agency built its first interplanetary probe with the ambitious objective of making a close flyby. France had hoped to fly an ultraviolet telescope on an American probe to fly to comet Halley, but in autumn 1979, toward the end of the parsimonious Carter presidency, NASA finally ran out of funds for a comet Halley mission. At the Corsica meeting, the Institute of Space Research (IKI) of the Soviet Academy of Sciences made the offer to the French to carry their telescope on the same Venus orbiter that would drop the French balloon. Comet Halley would come to within 40m km of Venus, much closer than it would to Earth. From a vantage point in Venus orbit, the French telescope would obtain good images of comet Haley.

Afterwards, the Institute of Space Research reflected on the mission: 40m km was still quite a distant view. Was there a way of getting the spacecraft and its telescope closer to Halley? Following a suggestion by Vladimir Kurt, examination of celestial mechanics by A.A. Sukhanov suggested a daring possibility. It would be possible, by careful choice of launch date, to send a probe to Venus, drop a smaller balloon and a lander and then change the course of the mother ship to intercept comet Halley. The requirement for a course alteration meant that the large balloon would now be a small balloon. The mission went through a number of further permutations before it was finally redesigned and the new arrangements settled.

VEGA: THE MISSION

At this stage, the project acquired a new name: the VEGA mission. This title came from the VE of Venus and the HA of Halley ('H' and 'G' are similar sounds in Russian). The in-house designator also used the term 5VK, following its OKB-1 heritage (some texts have also called it VEHA and the terms Venera 17 and 18 have also been occasionally used). This new name was announced in April 1982. The French, though, went off in a sulk about the downscaling of the balloon and declined to participate in a small-balloon project, which was now repatriated and had the letters 'CCCP' imprinted on it in large letters so that all the Venusians could be certain of the country of origin. Sagdeev cleverly sold the mission to the Soviet military on the basis that the mission would be a vivid illustration of the Soviet Union's capacity to deliver multiple, independently targeted warheads – albeit on another planet – and that this would really frighten the Americans. One incredulous military man told him privately that he didn't believe a word of this, but thought it was a great project and would support it anyway.

With the help of Boris Petrov, head of Intercosmos, an appeal was launched by IKI to invite international participation in the mission, the first time the Soviet Union had managed a mission in this way: 120 kg was available for international experiments. Thirteen European countries responded, even the huffy French. The telescope to follow the comet was built by Czech engineers, while part of the navigation system was built in Hungary. Now the ground rules changed. Instead of handing over their equipment months – even years – in advance and never seeing it again, foreign scientists were fully involved in the project before, during and after the mission. They visited Moscow frequently (collaborative meetings were held in many locations) and had a free run of the facilities and laboratories while they were there, unimaginable only a few years earlier.

The Cold War prevented the Americans from participating formally, even though it gave them an opportunity to make good their own lack of a probe to Halley. The 1972 Treaty on Space Cooperation which had inaugurated the Apollo Soyuz Test Project provided for cooperation in a number of areas, including deep space exploration, for five years. The treaty was renewed in 1977 but President Reagan had let it expire in 1982, so there was no forum where the United States and Soviet Union could formally work together. But both countries had agreements to cooperate with France, and here CNES investigator Jacques Blamont persuaded no fewer than 20 observatories worldwide to participate in the tracking of his VEGA balloons. Scientists formed an inter-agency consultative group, comprising European, Japanese, Soviet and American scientists, effectively by-passing the Reagan boycott of cooperation

VEGA spacecraft

with the USSR. The Jet Propulsion Laboratory did agree to track the mission with a view to improve targeting toward the comet, both for VEGA and the subsequent European *Giotto* mission.

In the event, following an informal approach, the University of Chicago did participate privately and fly an instrument to measure dust and cloud particles, with the assistance of NASA's Ames research centre. Because of restrictions on scientific contact, this had to be portrayed as a private venture. The scientist involved was quizzed by the military as to the perils of the Soviet Union copying advanced American technology, possibly tipping the balance in the Cold War. He assured them that he had bought only old components from the local radio store. 'Let them copy this,' he chuckled, 'it will set them back years!'

VEGA 1 and 2 were duly launched on 15th and 21st December 1984. The level of international collaboration prompted an unusual degree of openness for the USSR, and for the first time pictures of a Proton launching were published. Winter snow had

VEGA launch

not yet fallen at Baikonour. At take-off, the launcher was temporarily obscured by billowing clouds of nitric smoke. The censors had the last word, for the pictures showed only the first moments of launch, not staging. Not long afterwards, Mikhail Gorbachev came to power to usher in the brief, heroic age of *glasnost* and *perestroika*.

VEGA: SPACECRAFT AND INSTRUMENTS

The profile of the VEGAs toward Venus was different from anything done before. Instead of taking a straight path in toward Venus' orbit around the Sun, two VEGAs would be launched a month ahead of the normal launch window, immediately curve sharply inside Venus' orbit and then intercept the planet from the sun's other side a month after the normal arrival window. The fuel load would be increased from the normal 245 kg to 590 kg. For the mission, a number of modifications were made to

Vladimir Perminov

the spacecraft. The power demands of the mission required an extra set of solar panels on either side, like Venera 15 and 16, bringing the total area to $10\,m^2$. A 5 m long magnetometer boom was installed. A dust protection screen 0.4 mm thick was installed to protect the spacecraft.

For the first time, a Soviet space probe carried a manoeuvrable camera platform, one weighing 82 kg. Instead of having the space probe point at the target, the moving platform, built in Czechoslovakia with Soviet and Hungarian assistance, would swivel around and take pictures and was designed to cope with a closing speed with the comet of 80 km/sec. The television system had a resolution of 150 m from 10,000 km, using a high-resolution panoramic camera and light filters for pseudocolour images. To protect the spacecraft during comet encounter, green anti-dust screens were added to the spacecraft, considerably changing its appearance. A lightweight structure was added, to shake off particles attaching themselves to the station and forming their own atmosphere or ionosphere. A computer was provided with 816 bit memory and transmission rates were 3,072 bits/sec (routine, recorded and dumped data) or 65 kbps (real time, during encounter). The radio-electronic systems were again developed by Alexei Bogomolov's Moscow Engineering Institute. Technical director of the VEGA missions was Vladimir Perminov.

The balloon's payload was a cylinder with a conical antenna at the end. Inside were batteries designed for 50 hours, transmission systems and the scientific instruments. Readings were taken every 75 sec and transmitted every half-hour on 1.67 GHz at a meagre 4 kbps. The balloon had 2 kg of helium for inflation. The gondola was 1.2 m long, tapering to a cone, 14 cm in diameter, weighing

Vyacheslav Linkin

6.9 kg with a 4.5 W transmitter sending on 1.667 GHz. Total weight of each balloon was 21 kg. They were painted white to ward off the worst effects of the corrosive atmosphere. The gondola carried the scientific instruments, 1 kg of battery and a transmitter sending 4 bit/sec bursts from the 1,024 bit memory every 30 min for 270 sec bursts. The balloon designer was Vyacheslav Linkin (b. 1937) of the Institute of Space Research, a veteran of earlier Venus missions and an expert on the atmosphere of the planet.

VEGA: mother craft experiments
Television system (USSR, France, Hungary).
Three-channel spectrometer (USSR, Bulgaria, France), to analyze the chemical composition of the comet's matter in visible, ultraviolet and infrared bands.
Particle impact mass spectrometer (USSR, Germany), to examine mass of solid dust particles.
Neutral gas mass spectrometer (Germany), to determine molecular composition of comet's gas.
Dust particle detector (USSR), to measure the intensity of the dust particles.
Charged particle analyzer (USSR, Hungary), to characterize electrons and ions in the comet's plasma.
Plasma wave analyzer (USSR, Poland France, Czechoslovakia) for near-comet plasma.
Magnetometer (Austria), to measure comet's own magnetic field.

VEGA: lander experiments

Name	Objective
Meteo (with France)	Measure weather, pressure and temperature below 110 km
IFP	Measure aerosols at 40–50 km
ISAV (with France)	Nephelometer/scatterometer/spectrometer to measure composition of atmospheric gases 50 km to surface
Malachite (with France)	Mass spectrometer to measure composition of atmosphere at 40–50 km
Sigma 3 gas chromatograph	Measure composition of gases at 35–50 km
VM-4 hydrometer	Measure water vapour at 35–50 km
G515-SCV	Gamma ray spectrometer to detect uranium, thorium, potassium on surface
BDRP-AM25	X-ray spectrometer to detect silicon, aluminium, iron, magnesium on surface

VEGA: balloon experiments

Instrument	Countries	Aim
Pressure sensor	USSR, France	Measure pressure
Temperature sensor	USSR, France	Measure temperature
Wind meter	France	Measure wind
Nephelometer	USSR, France	Measure clouds and aerosols
Photometer	USA, France	Detect lightning
Position indicator	USA, France, USSR	Measure location and drift

VEGA AT VENUS

VEGA 1 corrected its course within days, on 20th December. Six months later, on 9th June 1985, VEGA 1 arrived at 500,000 km over Venus. The VEGA landers were targeted at a completely different part of Venus, the *Mermaid Plains* and *Aphrodite Mountains*, a hemisphere away from the traditional landing sites. In a modification to the lander, a circular ring was installed beneath the aerobrake so as to stop it spinning during descent.

The VEGA 1 lander was released at 39,000 km. VEGA 1's mass spectrometer was turned on at 64 km when the parachute was opened and it at once began to detect cloud particles, sulphur dioxide and chlorine. The X-ray fluorescent spectrometer was switched on 25 km up. The lander returned data during its descent and touched down on the *Mermaid Plains* lowlands at latitude 7°11'N, longitude 177°48'. The surface temperature was 452°C and the pressure 86 atmospheres (later corrected to 93). The mother craft flew past the planet at 8,890 km.

VEGA assembly conference

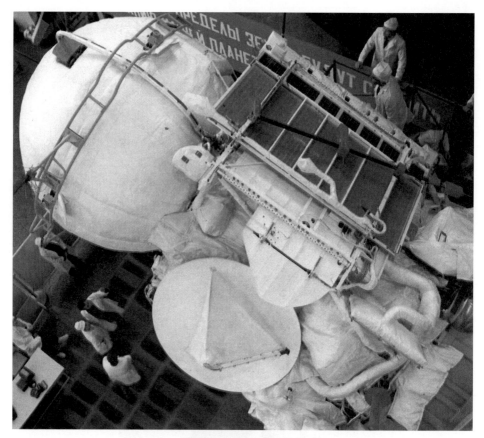

VEGA with lander shell

Few scientific results from the surface were published. Aware of the pressure for results following the VEGA 1 descent, IKI convened a press conference only three hours after touchdown. The preliminary results: the atmosphere contained sulphur dioxide, 0.01% water vapour and there were five distinct layers. Then IKI announced that VEGA 1 lander data could not be released until the comparable VEGA 2 lander results were available a few days later, which should have alerted suspicious minds to the fact that something had gone wrong. When the second VEGA did come down, Soviet reports then referred generically to the results of the 'VEGA landers', making no distinction between the two. Another warning sign. Trying to explain this, alert writers picked up rumours at the time that the drilling sequence started prematurely when the probe was still 16 km above the surface some 15 min before touchdown [9].

Years later, we found out what really happened. The electrical anomaly encountered by Venera 11–14, whilst it had no effect on VEGA 2, had a major impact on VEGA 1, triggering off the landing programme at an altitude of 18 km. Twenty years later, reports are now more scrupulous in attributing the surface reports to VEGA 2 only. The gamma ray device on the VEGA 1 lander did work and tested the surface,

VEGA international collaboration

but there was no drilling by VEGA 1, so the precious lowland–highland comparison could not now take place. *Mermaid Plains* were low plains from which no sample had been taken before, so this was a big scientific loss.

The absence of surface pictures created speculation, as it had with Venera 11–12, that the cameras had again failed. The Soviet authorities put forward, as an explanation, the fact that the landing was at night, but this was unconvincing since floodlights could have been carried, as they had been on Venera 9–10. The reality was that a decision had been taken not to carry cameras, but to fly other scientific experiments instead and this is confirmed by the final pictures of the landers, for cameras were not fitted.

Despite its difficulties, during its descent VEGA 1 took many samples of the atmosphere, detecting suphur, chlorine and phosphorous, with 1 mg/cm^3 of sulphuric acid in the range 48–63 km. Cloud particles were measured as smaller than 1 μm, like an Earthly fog, except more corrosive. It went through two cloud layers: the first 50–58 km, with the lower layer descending to 35 km. Surface temperature was inferred at 468°C and pressure 95 atmospheres. An improved hydrometer was carried to refine the water vapour data from Venera 4, finding that water vapour was only 0.15% at upper altitudes and much less lower down.

VEGA 2 dropped its lander on 15th June, carrying out a deflection manoeuvre and passing the planet at 24,500 km. During its descent, VEGA 2 continued the work begun by Venera 12 in making flourescent spectrometer analysis of the components in the clouds, finding not only sulphur, chlorine and iron but also, in the lower cloud layer, phosphorous (6 mg/m^3). The malachite experiment collected cloud particles and

VEGA landers

droplets and put them in a box for analysis, confirming Venera 12's identification of sulphuric acid, as well as chlorine, iron and phosphorous.

VEGA 2's landing module came down on the dark side of Venus on 16th June. The mother craft flew past at 8,030 km. For VEGA 2, the landing point was latitude 6°27'S, 181°5' in *Aphrodite Mountains*. *Aphrodite Mountains* comprise a mixture of lowland plains to foothill mountains up to 4 km above datum to *Atla Mountains*. VEGA 2 was targeted at an upland slope at an elevation of 1.8 km. Surface temperature was 460°C and pressure 90 atmospheres. Repeating the operations of Venera 13 and 14, the surface rocks were drilled. The drilling operations and transport to the container took 172 sec. The sample was then irradiated by fluorescent irradiation (iron-55 and plutonium-238) to detect their elements and their proportion. The VEGA rocks were found to be like Earthly basalt, not unlike that found in the northern Apennines in Italy. They were rich in aluminium and silicon, but poor in iron and magnesium. The rocks were like the anorthosite samples found on the moon and some were anorthosite–troctolite, a lunar highland rock rarely found on Earth. There was evidence that water had been present in the rocks when they melted on Venus' mantle, many geological years ago. According to the Academy of Sciences analyst, V.L. Barsukov: 'It can be concluded that a more water-saturated atmosphere and even hydrosphere may have existed in Venus geologic past.' Sulphur at the VEGA 2 landing site was higher, at between 2% and 5%, much more than Venera 13's 0.65%, suggesting that the VEGA 2 rocks were the oldest yet found. VEGA 2 was the first to combine a gamma ray spectrometer and X-ray flourescent spectrometer, the principal investigator being Yuri Surkov. But the mass spectrometer failed.

Descent data were transmitted from VEGA 1 for 62 min before its landing and VEGA 2 for 60 min during the descent. Reports on the transmission times from the surface are confused. Most give surface times for VEGA 1 of 56 min and VEGA 2 57 min, but some give surface times of as little as 22 min.

THE BALLOON JOURNEYS

The novel stage of the Venus part of the mission was the deployment of the first balloons in the atmosphere of the planet. Instrumentation for the balloon had been built in Austria, Hungary, the GDR, Bulgaria, Poland, West Germany, Czechoslovakia and even France although it had fallen out of the original project. The balloon was 3.4 m in diameter, and suspended 12 m below was a gondola. The VEGA 1 balloon deployed on the night side of Venus on 10th June 1985, about 11,000 km north of the equator. The snow-white VEGA 1 balloon was released at an altitude of 54 km. The aim was to deploy the balloons at an altitude where the cloud was thickest. VEGA 2's deployed a similar distance south of the equator.

The first word that the balloons had succeeded reached the control room in IKI on the telephone line from California, from where JPL scientists told their colleagues in Moscow that a balloon signal had just been picked up from the NASA dish in Australia. There, as the signals came in, the computer screens came alive and applause broke out. At IKI in Moscow where – due to the rotation of the Earth – the signals came in later, the mood was more reserved [10].

The balloons had an amazing journey, bobbing in the planet's atmosphere at around 55 km. They entered on the night side, drifting in toward the day side. The balloons were tracked by radio telescopes in Texas (Fort Davis), Iowa (North Liberty), West Virginia (Greenbank), Germany (Effelsberg), Sweden (Onsala), Spain (Madrid), South Africa (Hartebeesthoek), Canada (Penticton), Australia (Canberra), Brazil (Itepatinga), Britain (Jodrell Bank) and California (Goldstone and Owens Valley), including the two 70 m Soviet dishes in Yevpatoria and Ussuriisk, the 64 m dish at Bear's Lake and the smaller dishes in Ulan Ude (25 m) and Simeiz

VEGA aerostats in flight

and Pushkino (both 22 m), as well as the largest dish in the world, the 305 m crater dish at Arecibo, Puerto Rico.

As it turned out, the release altitude was also where the atmosphere was most turbulent. The VEGA 1 balloon was caught in turbulent air currents, whirled around at 240 km/hr and tossed up and down in air currents. The VEGA 2 balloon also circled at 54 km above the planet and recorded 'vortices the size of hurricanes'. VEGA 2's balloon whirled about through hurricanes, at one stage falling 3 km in 30 min and bursting after a day. Vertical winds were often 1 m/sec and sometimes up to 3 m/sec. No one had expected the atmosphere to be so stormy.

Dishes at Yevpatoria

VEGA gondola

Both balloons found themselves reguarly bouncing 200–300 m up and down, with downgusts of 1 m/sec, sometimes reaching 3 m/sec. The balloons found that the clouds travelled in gales at up to 60–70 m/sec. They themselves travelled at around 55 km/hr. Over the *Aphrodite Mountains*, where the strongest currents were encountered, the balloon plunged 2,400 m in an air pocket.

The gondola's instruments reported back on temperature, winds, light and the composition of the atmosphere, sending back 69 bursts of signals. After a 46 hr journey of 9,000 km through the darkness, the VEGA 1 aerostat reached the daylight side of the planet, where the sun heated its envelope and it burst. The batteries were built to last two days and it is possible that the batteries ran out before the balloons burst.

Venus atmospheric science from the VEGAs
The clouds of Venus are a thin fog of sulphuric acid with sulphur, chlorine and phosphorous. Sulphuric acid comprises $1.5 \, \mu g/m^3$.
Density of sulphuric acid is $1 \, \mu g/m^3$ at altitudes of 48–63 km.
The sulphuric acid may help to account for its yellow tinge.
Clouds retain some of their light even at nighttime.
Clouds are thickest (i.e., highest particle counts) at 50–58 km.
Water content was between 0.01 and 0.02% at between 25 km and 30 km and between 0.1% and 0.2% at 55 km to 60 km altitude.
Atmosphere is turbulent, even violent.

BREAKING THE CURSE OF SUSLOV: VEGA AT COMET HALLEY

Now, in July 1985, the mother craft altered course to pursue comet Halley in a 708m km chase. Benefitting from a gravity assist by Venus, their orbits now took them back out across the solar system, passing by the circle of Earth to the region between Earth and Mars, an elliptical arc that brought them back in toward the sun again as comet Halley rushed in for its solar encounter. In January of the new year, 165m km away from Earth, the instrumentation for the comet interception was turned on, in good time for closest approach in the first and second weeks of March. On 10th February 1986, the final course correction was made, on the 14th the scan platform locked on the 150m km distant comet and on 4th March the first images were received from a rapidly closing distance of 14m km. A day later, it was only 7m km out and another set of pictures came in. Then, 170m km away from Earth, close encounter took place early on 6th March 1986 when VEGA 1 entered the comet's corona or tail.

VEGA 1 closed in at an interception speed of 79.2 km/sec, a hundred times faster than a bullet. Cold War or not, American television covered the VEGA interception of comet Halley live, taking a feed direct from the IKI control room with commentaries by renowned astronomer Carl Sagan and IKI director Roald Sagdeev. Two hours before closest approach, VEGA 1 switched to high-rate telemetry. Now, as the psychedelic false-colour images of Halley filled the huge wall screen, covered with mathematical details of the interception alongside in cyrillic, the international scientists and media burst into applause. In real time, the interception of a comet! For the first time, Western correspondents were in a Soviet mission control room for live results from a deep space mission. *Glasnost* had at last broken the curse of Suslov. The room was filled with computers, display screens, computer disks, printouts, VEGA models and scientists conferring in small groups.

VEGA 1 passed the comet at a distance of 8,890 km travelling at 79.2 km/sec, taking over 500 images in the next three hours through different colour filters of the comet's oblong core and of dust jets spewing material out. VEGA 1 was damaged during flyby, losing 40% power from its solar panels. Final images were taken on 7th–8th March.

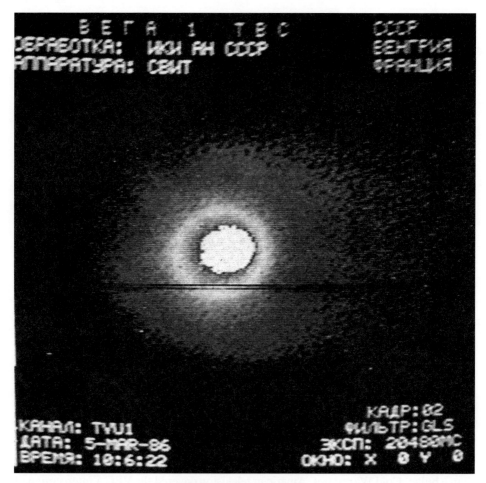

VEGA closes in on Halley

Now for VEGA 2: its interception path was perfect and the final course correction on 17th February was cancelled. The first images, a hundred of them, came in from a 14m km closing distance on 7th March. A couple of days later, VEGA 2 passed the comet at a distance of 8,030 km at 76.8 km/sec, taking images of the south pole of the comet's nucleus. There was a worrying moment half an hour before VEGA 2's close approach, for the computer guidance system failed. Thankfully, it was able to switch over quickly to the backup system. The dust around the comet caused power to fall by 80%. Again, blue, white and red images of the potato-shaped comet spewing vapour filled the control rooms in IKI. Final shots were taken of the receding comet on 10th and 11th March, 7m and 14m km distant. VEGA 2 returned 700 images.

Data from the two probes were very similar and the flybys executed with equal precision, leaving Soviet scientists and their international colleagues elated. These

VEGA mission control

were not the closest passes, for the European space probe *Giotto* made a daring, arguably dangerously close approach of only 500 km.

There was some discussion of retargeting the VEGA probes toward the asteroid Adonis, but this did not happen. Their missions effectively concluded in March 1986.

THE INTERCEPTION OF HALLEY: SCIENTIFIC RESULTS

The scientific information returned from Halley by the VEGAs was enormous. Between them, the two VEGAs sent back over 1,500 pictures of the comet's glowing gases, the multispectral images giving the blob of the comet a psychedelic red and yellow gassy surround against a blue and purple sky. Halley's Comet, the VEGA had determined, was an oblong monolith of black, clathrate, carbon dioxide ice 14 km long and 7 km wide from which millions of tonnes of vapours were escaping, with a surface temperature of 100°C but an internal one of 100,000°C. The gas flows comprised hydrogen, oxygen, carbon, carbon monoxide, hydroxyl and cyan. The comet was covered with a porous crust and found to contain elements familiar in some stony or metallic meteorites. The VEGAs took a chemical analysis of the dust particles blown away, some of them hundredths of a micrometer in size. These particles

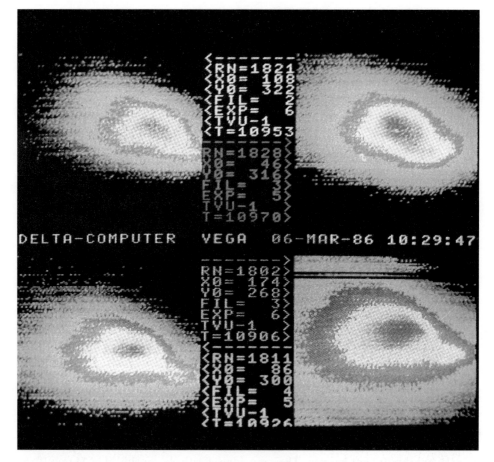

VEGA 1 at Halley

contained mainly carbon but also sodium, magnesium, carbon, calcium, iron, metals, nitrogen, silicates, oxygen and hydrogen. There was chlathrate ice (carbon dioxide ice). The mass of the comet was estimated at 300bn tonnes. A million tonnes escaped every 24 hours at perihelion passage. Halley was turning around itself every 53 hr and had a dark surface, reflecting only 5% of light falling on it, possibly with a black porous skin 1 cm thick.

Russian scientists were now able to model the comet. It was a conglomerate of ice with a self-renewing thin layer of black porous matter of low heat conductivity. From time to time, the inside of the comet would evaporate and burst open on its surface, creating active zones. Carbon dioxide and ice would break through and solid particles would be carried away, the comet losing tonnes of material a day, but quickly rebuilding itself. As gas escaped from the comet at 1 km/sec, the solar wind

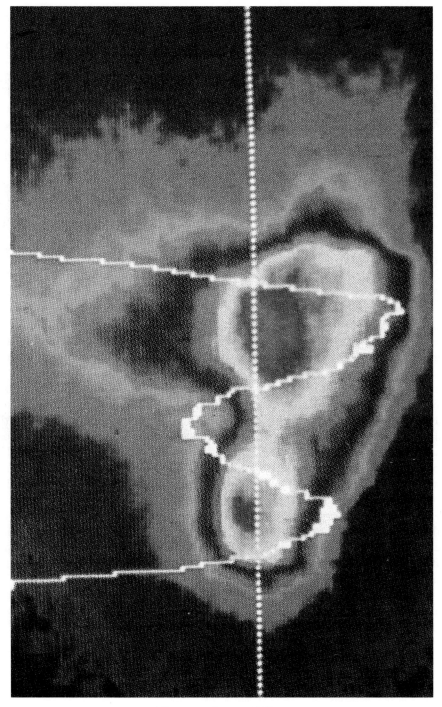

The core of comet Halley

would ionize the evaporated gases, causing the long, luminous cometary tail and a plasma cloud 1m km wide and 25 times greater than Earth's magnetosphere.

The main chemicals of the comet were found to be water vapour and carbon dioxide, with atomic hydrogen, oxygen, carbon and molecular carbon monoxide and dioxide, hydroxide and cyan. As many as 2,000 particles were studied by the dust impact mass spectrometer and they varied a lot in composition, some being metallic, with sodium, magnesium, calcium and iron. Some had silicate, some had oxygen and hydrogen (water elements), while others had carbon. Some were very small, only microns across. From all of this, the scientists concluded that cometary nuclei were probably formed near the sun between Jupiter and Neptune and subsequently hurled away.

HIGH SUMMER

VEGA marked the climax of the high summer of the Soviet interplanetary programme. Following the disappointment of the great Mars fleet and the ensuing 'war of the worlds', Venus had become the prime target of Soviet interplanetary interest. The years 1975–1986 had seen unbroken success with missions of ever greater complexity. In a dramatic contrast to the earlier years, there had not been a single launch failure. During this period, eight landers had descended to the surface of Venus, transmitting information during the descent, on and after landing, enabling scientists to build up a picture of the atmosphere, clouds, temperatures and pressures of the planet. Camera equipment had imaged the surface, probed the density of the soil and brought it on board for chemical analysis in harsh and unforgiving conditions. The four orbiters had built up a picture of the atmosphere from above. The two radar mappers had characterized the immensely varied and complicated surface of the planet. The balloons had given a bird's eye view of the cloud systems. The VEGAs were the final *tour de force*, as sophisticated a mission as planners could ever devise, dropping landers and balloons and then flying across the solar system for the Halley interception where the rendezvous had been carried out in the full glare of live television publicity. A familiar Western criticism of the Soviet space programme was that, whatever its successes, it was crude and unsophisticated. No one could say that now.

What did they all achieve? This may be a good point to summarize the results achieved by all the landers going back to Venera 7 fifteen years earlier. Thanks to good targeting, the landers came down in a broad range of sites:

Old upland rolling plains	Venera 8, 13, VEGA 1
Flat lowlands	Venera 14
Young volcanic structures	Venera 9, 10
High mountain massif slopes	VEGA 2

Lander transmission times amounted to several hours. These represent the period for which signals were received, the craft either succumbing to the pressure and heat or, alternatively, the mother craft moving out of transmission range.

Lander transmission times

Venera 7	23 min
Venera 8	63 min
Venera 9	56 min
Venera 10	66 min
Venera 11	95 min
Venera 12	110 min
Venera 13	127 min
Venera 14	57 min
VEGA 1	56 min
VEGA 2	57 min

Across the Veneras, surface temperature ranged from 452°C (the lowest) to 474°C (the highest). Pressure varied from 85 atmospheres to 94 atmospheres, largely a function of where the probes landed in respect of the reference level.

The probes identified vulcanism as the main feature shaping the surface, with no evidence of plate tectonics. The rocks were rounded, the most likely reason being wind erosion, confirmed by wind speed measurements of 0.4–1.3 m/sec. The strength of Venusian rock varied, being weak at the Venera 9 site, strongest at Venera 10. Venera 13's had the consistency of heavy clay combined with compacted fine sand while Venera 14 was more like foam concrete. Chemical analysis of the rocks revealed the main elements to be silicon (typically around 45%), aluminium (about 16%), magnesium (11%) and iron (8%). This was the overall picture, but there were variations between sites. Venera 8's was quite different from the others, with 4% potassium (less than 0.45% elsewhere), high uranium (2.2%, less than 0.69% elsewhere) and high thorium (6.5%, less than 3.65% elsewhere).

What about the atmosphere? Knowledge of the Venusian atmosphere was obtained by a mixture of instruments on the descending landers or the orbiters.

Russian spacecraft orbiting Venus

22 Oct 1975	Venera 9	1,510–112,200 km, 48 hr 18 min, 34.17°
25 Oct 1975	Venera 10	1,620–113,900 km, 49 hr 23 min, 29.5°
10 Oct 1983	Venera 15	1,000–65,000 km, 1,440 min, 87.5°
14 Oct 1983	Venera 16	1,000–65,000 km, 1,440 min, 87.5°

The clouds could be broken into a number of layers, though below 50 km they were really more like mist or haze than our clouds.

Top layer	64 km
High density	50–60 km
Moderate density	32–49 km
Mild	18–31 km
None	Below 18 km

The composition of the atmosphere varied according to height, water content being in the clouds while carbon dioxide and oxygen were higher above the clouds. For example, sulphur and chlorine were most evident from 46 km to 63 km. Venera 12 measured the composition with some detail and is probably the best guide, finding the main elements to be carbon dioxide, 97%, with nitrogen 2–3% and then a series of minor elements.

Most solar energy was absorbed at high altitudes, over 50 km. Only 1% of solar energy reached the surface, but this was sufficient to maintain high temperatures in the low atmosphere and contribute to a greenhouse effect. The original 'greenhouse effect' theory of Sagan appeared to be confirmed, a grim warning of the ultimate effects of climate change on Earth itself.

By the late 1980s, not only were the physical details of Venus apparent, but scientists were beginning to interpret the data to form a picture of Venus' origin, development and evolution. A basic problem remained, though. Before the space age, it had been assumed that planets of similar size, formed in a similar point in space with a similar chemical composition and early history should follow a similar evolutionary path – hence the images of swamps on Venus. But, at some point Venus and Earth diverged and this remained a challenge for planetologists. But, at least with the Venera and American data, they had some facts to work on. With the Venera and VEGA landings, the radar maps and now the deployment of the balloons, most of the key information about Venus's atmosphere, clouds, temperature, pressure, surface, rocks and composition was now obtained. The transformation in the level of knowledge between 1961, when the programme started, and now, could not have been greater.

Did this mean that all the questions about Venus had been answered? Far from it. When Mikhail Marov and David Grinspoon reviewed the state of knowledge of Venus in 1998, it was obvious that the missions had told us *What Venus is now* but not *What Venus was* or, more importantly, *How did Venus become like this?* [11]. The Venera missions did not, could not, tell us how Venus evolved, how it became so hot or what had happened to its atmosphere, although there was now an abundant bed of knowledge upon which to construct different theories, possibilities and explanations. With the Earth facing climate change, these issues acquired a contemporary relevance and were no longer just of scientific or academic interest. There was one consolation, though, for those artists who drew those moist picture of Venus's swamps, jungle canopies and blue plants. They were accurate pictures of Venus after all – but several billion years ago, not in the 1950s.

Scientific knowledge of the two planets closest to Earth had now been through considerable evolution, if we look at the different paradigms of their evolution through the ages.

	Venus	Mars
1700s		
1800s		Increasingly Earth-like
1880s		Canali (Schiaparelli)
1890s		Cold, dying civilization (Lowell)
1900s		Belligerent civilization (Wells)
		Benign civilization (Boguslavski)
1930s	Oceans, swamps, vegetation	Lichens: cold, dry, windy
1950s	Possibly hot?	
1960s	Hot, high pressures	Lifeless desert, like moon
1970s	Greenhouse effect	Cold, dry but past water
1980s	*Why so different from us?*	Warm, wet, early Mars

See: Jeffrey S. Kargel: *Mars – a warmer, wetter planet.* Springer/Praxis (2004)

Three more Venera-type probes were constructed. Later, in 1985 it was decided that the interplanetary programme should be refocused around Mars and to terminate Venus exploration for the time being [12]. The present model had probably been flown to its limits. What to do with the three remaining spacecraft? In all cases, it was decided to remove the landers and convert the main part of the spacecraft to house a telescope. The first was flown as an astrophysical observatory, *Astron*, in 1983. The second was completed and flown as an observatory *Granat*: it was especially successful, operating for ten years. The third was designated *Lomonosov*, a star mapper, intended to plot the position of over 50,000 stars with great accuracy. Its launching was set for 1992, but the conversion of the hull for this purpose was never completed.

The 1980s were characterized by considerable optimism about the Soviet space future. The broad investment across the programme – that was a feature of the Brezhnev period – reached its peak then and the Soviet Union was launching as many as a hundred spacecraft a year, or two a week. February 1986 saw the launch of what became the greatest of all the space stations, *Mir*. In May 1987, the ailing Valentin Glushko, the chief designer who – when he was a little boy – had corresponded with Konstantin Tsiolkovsky, oversaw the launching of the most powerful rocket ever constructed, the ultramodern *Energiya*. Soviet space industry now operated on so vast and successful a scale as to attract the compliment of a special profile in the United States' *National Geographic* magazine, the first time it had covered the Soviet space programme. Commemorating 30 years since Sputnik, *Time* magazine ran a cover story *Moscow takes the lead*, with the story titled *Surging ahead – Soviets overtake the US as the No. 1 spacefaring nation.*

REFERENCES

[1] Barsukov, V.L.: *Basic results of Venus studies by VEGA landers.* Institute of Space Research, Moscow, 1987.

[2] Breus, Tamara: *Venus – the only non-magnetic planet with a magnetic tail.* Institute for Space Research, Moscow, undated.

Astron

[3] Burchitt, Wilfred and Purdy, Anthony: *Gagarin*. Panther, London, 1961.

[4] Surkov, Yuri: *Exploration of terrestrial planets from spacecraft – instrumentation, investigation, interpretation*, 2nd edition. Wiley/Praxis, Chichester, UK, 1997.

[5] Once again, much of our knowledge of the instrumentation on these Venus probes comes from Mitchell, Don P.:
  - Soviet interplanetary propulsion systems;
  - Inventing the interplanetary probe;
  - Soviet space cameras;
  - Soviet telemetry systems;
  - Remote scientific sensors;

– Biographies;
– Plumbing the atmosphere of Venus;
– Drilling into the surface of Venus;
– Radio science and Venus;
– The Venus Halley missions, *http://www.mentallandscape.com*

[6] Basilevsky, Alexander: The planet next door. *Sky and Telescope*, April 1989.

[7] Lemonick, Michael D.: Surging ahead. *Time*, 5th October 1987.

[8] For example, Kuzmin, Ruslan and Skrypnik, Gerard: A unique map of Venus. Novosti Press Agency Soviet Science and Technology *Almanac*, 1987.

[9] Woods, Dave: Probes to the planets, in Kenneth Gatland (ed.): *Illustrated encyclopedia of space technology*. Salamander, London, 1989.

[10] Soviet space odyssey. *Sky and Telescope*, October 1985.

[11] Marov, Mikhail Y. and Grinspoon, David H.: *The planet Venus*. Yale University Press, New Haven, CT, 1998

[12] Huntress, W.T., Moroz, V.I. and Shevalev, I.L.: Lunar and robotic exploration missions in the 20th century. *Space Science Review*, vol. 107, 2003.

7

Phobos, crisis and decline

Show me a beaten man, for he is worth two unbeaten men.

– Old Russian proverb, quoted by Vladimir Perminov in
The difficult road to Mars [2]

VEGA 2 was the last Soviet or Russian Venus mission and remains so to the present day. In the 1980s, the Soviet Union and then Russia turned their attention back to Mars. By 1986, Soviet scientists were beginning to reach the limits of what could be achieved on Venus, although some further missions were sketched (Chapter 8). With the unbroken success of the Venus programme from 1975 to 1986, they had good reason to expect that their new efforts on Mars would be more successful than some of the previous missions.

Following the great Mars fleet, war had waged in Soviet scientific and political circles as to the focus of the interplanetary programme and whether and in what way they should compete directly with the United States. As Chapter 6 showed, this 'war of the worlds' was won decisively by the 'Venusians'. Or so they thought.

THE MARTIANS STRIKE BACK: PROJECTS 5NM, 5M

The 'Martians' might be down but they were not out. Now they presented plans to obtain rock samples from Mars. Not only would such a mission eclipse anything that the 'Venusians' could achieve, but it would be far beyond the reach of the old foe, the Americans.

The idea of obtaining samples from Mars was not a new one. The first, known as Project 5NM, was instigated at around the time of the recovery of the first samples of moon rock by Luna 16 in September 1970. Studies were carried out on the direction of

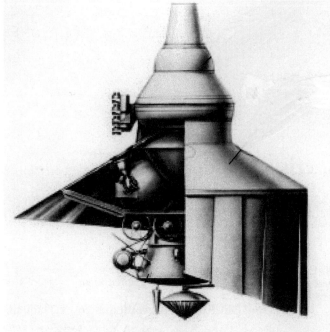

Project 4NM

Georgi Babakin and his starting point was the need for a rocket much more powerful than Proton. He proposed to use the N-1, returning to its original use for a Mars mission, though unmanned. The N-1 had then flown twice, both times unsuccessfully, but there was confidence that the rocket would become operational over the next number of years. Project 5NM was given a 1975 launch date.

Project 5NM drew from existing designs. The N-1 would send a 20-tonne spacecraft on the way to Mars. This would comprise:

- A 3,600 kg orbiter and relay station, based on the M-69 and M-71 design.
- A 16 tonne lander with a diameter of 6.5 m.
- In the top of the lander, a 750 kg return vehicle with a 15 kg cabin to contain 200 g of Martian soil.

The lander was a large spacecraft, its broad diameter being a function of the entry cone. Behind the cone were 30 petals, which would open to double the area of the cone to 11 m. The lander would enter the Martian atmosphere much like Mars 2 and 3. Once its speed had fallen to 200 m/sec, the cone would be jettisoned so as to make a rocket-only soft-landing, like the American Vikings. This was a huge spacecraft, five times the height of a human.

The lander would spend three days on the surface. Samples would be selected on the basis of panoramic pictures of the most interesting rocks. The ascent stage would then be launched, the precious samples on board. It would wait in a Martian orbit of

500 km, period 12 hr, for the next favourable alignment of Earth and Mars in ten months' time, when the capsule would be fired in the direction of Earth, where it would enter the Earth's atmosphere like Luna 16.

Project 5NM reached design drawings over the next two years. By 1973 it became apparent that it would not be ready for 1975, even if the N-1 was. Georgi Babakin felt that the chances of success were small, granted that so many features of the hardware were untested and the complex mission required faultless operation for three years. In a reversal of the normal role where reluctant ministers agreed to missions proposed by ambitious designers, Babakin lost confidence in the project. Minister Sergei Afanasayev, who was strongly in favour, reluctantly accepted that the project was unfeasible at this stage. By way of a precursor mission, designs were done of a Mars rover, or Marsokhod, very directly based on the successful moonrover, the Lunokhod. This was called mission Mars 4M. This would use the same aerodynamic cone and lander intended for project 5NM. Aerodynamic braking would apply and braking engines would be used for the last 3 km of the descent.

Being dependent on the N-1 for a launcher, 4NM was cancelled along with the N-1 project in May 1974. VNII Transmash actually made significant progress on rover design in support of the 4M mission. Four prototype rovers were built:

- 4GM, with four independent tracks;
- KhM, like PrOP-M, but fully scaled up, weighing 240 kg;
- KhM-SB, modelled on the lunar rover; and
- EOSASh-1, with six wheels and an articulated bogey system.

Despite the reorientation of the programme following the triumph of the 'Venusians' in 1974, an attempt was made to reverse their approach. The undoing, remaking and reversal of decisions was an endemic feature of the Soviet space programme, especially the lunar effort and the interplanetary programme was no exception. Now the 'Martians' found a new champion and the 'war of the worlds' was quickly renewed. He was the 80-year-old president of the Academy of Sciences, Alexander Vinogradov, the man who led the successful lunar soil recovery missions, Luna 16, 20 and 24 over 1970–6 and had been head of the Vernadsky Institute from 1947. His proposal was to refocus on Mars. In doing so, the Soviet space programme would revert to form, for an important objective was to pull off the mission ahead of the Americans.

In early 1975, not long into the apparent ascendancy of the 'Venusians', Alexander Vinogradov persuaded Dmitri Ustinov, Secretary of the Central Committee and Minister of Defence, to convene a special Saturday morning meeting in OKB Lavochkin. Vinogradov held out the prospect that recovered Mars samples would be the rosetta stone to the solar system. According to the leading 'Venusian', Roald Sagdeev, no one was prepared to admit to Ustinov that the project was well beyond the capacity of Soviet rocketry. There were no objections from the head of the space industry Sergei Afanasayev, a fan of the previous 5NM project, nor from the declining Mstislav Keldysh. Ustinov gave Lavochkin three years to prepare the mission. Lavochkin director Sergei Kryukov put his best efforts into the project, which had a 1980 launch date pencilled in.

Alexander Vinogradov

Because the N-1 rocket was no longer available, the much smaller Proton had to be used. This had several consequences. First, the spacecraft to be used had to be reduced in weight to 8,500 kg. Even still, two Proton rockets would be required to fire it out of Earth orbit. Moreover, a third Proton rocket would be required to send another spacecraft to Mars orbit to retrieve the samples later. This had quickly become a big enterprise. It was called Project 5M.

The 5M lander would descend to the Mars surface, using only retrorockets – no parachutes. The upper stage, of 2,000 kg, would fire out of the top of the lander to deliver the samples into Mars orbit for transfer and recovery. To prevent biological contamination, the return cabin would go into Earth orbit for retrieval where it would be scrutinized aboard an Earth-orbiting station before being brought down.

Project 5M required no fewer than three automatic dockings: one in Earth orbit at the start, one in Mars orbit and a third in Earth orbit on return. Several designers in Lavochkin felt that the project was so complicated as to have little chance of success. Kryukov persevered and, after some further evolutions, the design was finalized in January 1976. The weight of the spacecraft had risen to 9,335 kg, but some savings were made on the return leg. It was decided to sterilize the samples by heating them in Mars orbit, which meant that they would return direct to the Earth without biological decontamination in Earth orbit. Extreme design compromises were made. The 7.8 kg return craft would carry no parachute but a small radioactive device. Recovery helicopters would locate the samples – presumably impacted somewhere in the Siberian tundra – from their beeping radioactive beacons. The design was approved by the government and production of hardware began in 1977.

Eventually, the 'Venusians' struck back and the Institute of Space Research managed to persuade Sergei Afanasayev than the project was not feasible. Sometime in 1977 it was cancelled and the leading 'Martian', Alexander Vinogradov, was now dead. His place was taken, as director of the Vernadsky Institute, by a leading Venusian, Valeri Barsukov who held the post to 1992. In tennis terminology, this was game, set and match to the Venusians.

Project 5M

NEW CHIEF DESIGNER: VYACHESLAV KOVTUNENKO; NEW SPACECRAFT: UMVL

The cancellation of 5M sparked a crisis in Lavochkin. This was the first time any of its projects had ever been cancelled (the bureau had led a sheltered existence compared with some others). A dispirited Sergei Kryukov, who may have been skeptical at the start but had given the project his all, handed in his resignation and his place was taken by the deputy director of the Yuzhnoye design bureau Vyacheslav Kovtunenko. As for Kryukov, he moved to NPO Energiya where he worked as deputy chief designer to Valentin Glushko until 1982 when ill health forced him to retire. He died on 1st August 2005.

Vyacheslav Kovtuneko was born in Engels in southern Russia on 31st August 1921. He joined the army as soon as the Soviet Union was attacked by Germany in 1941, but was wounded early on and hospitalized for a lengthy period, not seeing fighting again. He enlisted in Leningrad University where he graduated in 1946, taking up his first job designing missiles in Moscow the same year. Seven years later, he moved to Dnepropetrovsk in the Ukraine, where he joined Mikhail Yangel's bureau

Vyacheslav Kovtunenko

and became the designer of the Cosmos and Tsyklon rockets and was subsequently responsible for the Intercosmos programme of small scientific satellites. He lectured students in the university there and wrote a book *Orbital spacecraft and aerodynamics*. His solid track record must have appealed to Lavochkin [1]. On the other hand, he was rated less well for interpersonal skills, where he had to contend with such heavy-weights in the space industry as Valentin Glushko, Nikolai Pilyugin and Mikhail Ryazansky [2].

Vyacheslav Kovtunenko's first decision was to get moving on a new design, a fifth generation of interplanetary spacecraft started by the 1MV, 2MV and 3MV series. Following the transistor problem with the great Mars fleet, such a decision had already been taken in principle by Sergei Afanasayev, but no technical progress had been made during Sergei Kryukov's time, presumably because his time was taken up by Project 5M. The new generation would be called the UMVL or Universal Mars Venus Luna spacecraft. This would serve a variety of purposes, from future Mars missions to lunar orbiters. Such a title deliberately attracted both the 'Venusians' and the 'Martians' and achieved an economy of scale. In the meantime, the Venera programme developed by Sergei Kryukov over 1973–8 would be brought to a con-clusion by the mid-1980s. Kovtunenko indeed did find it difficult to enlist the attention of 'the heavyweights' and, as a result, development of the UMVL was slow, taking the better part of ten years.

Like the Mars 2–7 spacecraft and new Veneras, the UMVL followed the bus concept, but there were important differences. The new spacecraft had a squat appearance, sitting on a series of large propulsion tanks and engines. The solar panels were now on the bottom, not the side. The high-gain antenna was on a hinge on the top. Scientific equipment was installed at a variety of points on the top of the frame rather than at a dedicated single point at the top (e.g., the landing cabin). Following Afanasayev's commitment to replace the old transistors, new digital electronic sys-tems were introduced. The UMVL was equipped with a computer able to store 4.8 GB of information.

UMVL

The spacecraft bus was propelled by the large combined braking/correction propulsion system (CBPS). This was a two-part system. The lower was the autonomous propulsion system (APS) of an engine with eight tanks – four of 730 mm and four of 1,020 mm for 3,000 kg of fuel. The APS was to be used for mid-course manoeuvres, braking into orbit and subsequent manoeuvres before being jettisoned. The CBPS second stage, which remained, had one central and four peripheral tanks, with 28 thrusters and was to be used for attitude control and minor manoeuvres.

PHOBOS: LAST OF THE SOVIET MARS PROBES

Despite the brief, unhappy foray of Project 5NM, the 'Venusians' remained otherwise ascendant. By the early 1980s they had achieved so much and the programme had gained so much in confidence that even they could no longer object to a return to Mars. The decision to return to Mars was a confluence of a number of factors: an exhaustion of what could be done with Venus, confidence that Mars missions could now be successful and the apparent loss of interest in Mars by the Americans [3].

Anticipating the spirit of *glasnost*, plans for a return to Mars were publicly announced in the mid-1980s. The surprise was that the primary target was not the planet itself, but its moons. This followed the non-competitive logic of the 'Venusians'. A mission to a Martian moon would not compete with the Americans, for they had never visited the Martian moons (though Mariner and the Vikings had photographed them from a distance), nor did they plan to do so. Indeed, the 'Venusians' in IKI felt that a mission to a Martian moon was a realistic, sensible substitute for the hopelessly over-ambitious Mars sample return.

Mars had two moons, both very small, written about by Jonathan Swift in 18th century Dublin though actually not discovered by astronomer Asaph Hall until 1877. Both orbited far out, right over the Martian equator. Phobos circled Mars every 7 hr 39 min about 6,000 km distant, while Deimos was much farther out, at 23,500 km, orbiting every 30 hr 18 min. Phobos was a 27 km diameter potato-shaped moonlet, with one crater, Stickney, named after Hall's wife's surname.

First plans for a mission to Phobos were announced on 14th November 1984, with the ambition of meeting the 1986 launch window to Mars, but by April the following year, this target had slipped to 1988. It was decided to construct two large six-tonne UMVL spacecraft and, following the success of VEGA, to invite international participation. Twelve countries duly responded. The cost of the project was estimated at 272m roubles, but the international part was valued at 60m roubles. In a significant change made by the Ministry of General Machine Building, the scientific supervising team was disbanded, leaving this work to be carried out alone by the contractors, the Lavochkin design bureau.

The Phobos mission went through a number of evolutions: from having the whole spacecraft land on Phobos, to a 20 m approach – with samples harpooned, dragged in and analyzed on board, to a close approach. The final decision was for a close approach, the dropping of landers and the use of ion beams to zap the surface, a beam being put together by Soviet, Czech, Bulgarian, Finnish and German scientists.

The new CBPS was critical to the mission. Two mid-course corrections were planned: between day 7 and 20 of the 200-day flight to the planet and on day 185 to 193, shortly before insertion into Mars orbit. The Phobos spacecraft would then find itself in a highly elliptical orbit around the planet Mars. Its next task was to move into an almost circular equatorial orbit about 350 km above the orbit of the moon Phobos, called an 'observation orbit'. Three tonnes of fuel were set aside for all these manoeuvres and once the observation orbit was achieved, the lower stage of the propulsion system would be dropped. This lengthy stage of the mission was a critical one, for the orbit of Phobos was not known precisely – or, at least, not precisely

Phobos hovering

enough for an approach and landing. Until then, measurements of the moon's orbit were accurate to within 150 km, but no more. Much more precise measurements were essential for an approach to be carried out with an economy of fuel or for a collision to be avoided.

The observation orbit would intercept Phobos at two points in its slightly elliptical trajectory. The second stage of the CBPS would then be used for the Phobos interception and, after that, to put the spacecraft into its final orbit around Mars. The rendezvous manoeuvre with Phobos was planned as follows:

First phase	Enter equatorial orbit 350 km farther out than Phobos, observe for a month.
Second phase	Synchronize orbit with Phobos, to within 35 km at end, two months.
Third phase	Interception.

The spaceship would rendezvous with Phobos, close in and drop a lander from a hover point 50 m above the moon. This comprised a long-term automated lander (LAL) and a hopper. The LAL would anchor itself to the low-gravity surface by harpoon, deploy solar panels and transmit for two months information from its spectrometer, seismometer and solar sensor. The LAL looked like a stool, with a

CBPS

tilted solar panel at the top. The seismometer would detect Phobos's own thermal expansion and contraction.

The 100 kg hopper, built by Alexander Kemurdzhian's VNII Transmash, the company that built the PrOP-M skid rover, looked like a bouncing ball, but flat at one end. It would be ejected out of the side of the spacecraft at closest approach and drop slowly to the surface. Once there, it would settle with the flat face down (if not, levers would turn it over till it did). Once settled, the spring would make it jump. The hopper had a spring to enable it to jump between 10 m and 40 m up to 20 m high and it carried an X-ray fluorescent spectrometer, magnetometer, penetrometer, dynamograph and gravimeter. Ten jumps were planned.

During the hover, the main spacecraft would bombard the surface 150 times with the 70 kg LIMA-D laser beam, shooting material 50 m high in ten-nanosecond bursts, evaporating material 1 mm down. Its spectrometer would measure the scattered ions. DION was a smaller, 18 kg, Soviet–French–Austrian krypton ion gun to measure secondary ions. High-resolution pictures by the Fregat system of cameras would image the surface down to 6 cm resolution.

After interception the mother craft would back away to 2 km distance. The 35 kg Soviet Grunt system would pulse-radar-map the surface to a depth of 2 m. The mother craft would then leave Phobos for a final equatorial Mars observation orbit to study its surface and atmosphere across a range of parameters, and continue in-flight experiments. Two scenarios were outlined for the final orbit: one a 7.6-hr orbit below the Phobos orbit, another much lower at 500 km [4].

All this required a high tracking capacity. The new transmitter and receiving stations, the P-2500 system – operating since the Venera 11 and 12 missions – was upgraded to receive signals at a rate of up to 131,000 bits/sec.

The largest-ever set of experiments for a Soviet interplanetary mission was then assembled. There were some differences as to the packages eventually carried on the two spacecraft; the intended list was not necessarily the final list and the same experiment sometimes received different titles and acronyms, so that assembling a

Phobos in assembly

definitive list has presented some problems. Both spacecraft came in overweight, so Phobos 1 lost the hopper, while Phobos 2 lost radiosounding, solar X-ray camera and neutron experiments.

Phobos experiments: in flight
Magnetometers.
Low-energy electron and ion spectrometer.
Solar wind spectrometer.
A proton/solar wind spectrometer.
Low-energy solar X-ray spectrometer.
Plasma wave analyzer.
X-ray telescope.
X-ray photometer.
Solar ultraviolet radiometer.
Gamma ray spectrometer.
Solar cosmic ray detector.
Solar photometer.
Gamma ray burst detector (same as Venera 11, 12).
Energetic charged particle spectrometer, SLED.
Solar telescope coronograph.

Phobos experiments: interception
Remote laser mass spectrometer (LIMA-D)
Remote mass analyzer of secondary ions (DION)

Phobos experiments: long-term autonomous lander
Television.
Seismometer.
Spectrometer.
Penetrator.
Telephotometer.

Phobos experiments: hopper
X-ray fluorescent spectrometer.
Magnetometer.
Penetrometer (PrOP-F).
Dynamograph.
Gravimeter.

Phobos experiments: Mars orbit
Scanning infrared radiometer TERMOSCAN.
Infrared spectrometer.
Gamma spectrometer.
Radio sounder.
Television unit (Fregat).
Neutron moisture meter.
Atmospheric spectrometer.

TERMOSCAN was a linear descendant of the pinpoint photometer cameras used on the earlier moon, Venus and Mars probes, except that it was cooled by liquid nitrogen and designed to obtain infrared images that would mark out areas of warmth in some detail, each image having $512 \times 3{,}100$ pixels. Typical resolution was 1.8 km. The designer was the long-time builder of observation systems, Arnold Selivanov.

Phobos was an ambitious mission by any standards, but it was based on the confidence generated from the experience of VEGA. America's leading planetary scientist Carl Sagan hailed the mission plan as novel, world class and very clever, but colleagues warned that its complexity was 'hair-raising' [5].

'WE CAN SHOOT THEM ALL LATER'

Phobos 1 rode an orange plume of flame into the night sky over Baikonour on 7th July 1988, witnessed by a large press corps, the international scientists involved and a visiting American military delegation. The censors had finally departed and this time staging could be followed high into the atmosphere. Two days later, ground control reported that all systems were deployed and functioning. A mid-course correction was executed on 16th July. Phobos 1 weighed 6,200 kg and became the biggest inter-planetary spacecraft ever launched. By chance, it left Earth during a magnetic storm which it was able to observe. During August, the Terek telescope on board Phobos 1 transmitted 140 pictures of the sun, including a 27th August picture of a plasma ejection the length of half a solar radius. The photometer registered a hundred bursts of hard gamma radiation.

Phobos 2 left Earth on 12th July, correcting its course on the 21st July. The mood in Moscow was upbeat and predictions were made of a Mars landing by cosmonauts between 2015 and 2017. The spacecraft were called Phobos 1 and 2, rather than Mars 8 or 9, to signal their distinct mission and, possibly, a fresh start to Martian exploration.

On 2nd September, ground control in Yevpatoria keyed up a standard command to Phobos 1 to turn on the gamma ray spectrometer. It was put about at the time that the unfortunate technician left one hyphen out of the series of keyed commands and the command which left Earth happened to be the end-of-mission command to close down all systems. It was an embarrassing, expensive mistake – and a foolish one to have designed computer commands in such a flawed manner.

In fact, the Phobos 1 failure was more complicated. There had been a political argument between Yevpatoria and Moscow as to which centre should control the mission. Moscow won, but, in compensation, Yevpatoria was given responsibility for checking all the commands that went up to Phobos. On 2nd September, Yevpatoria's checking equipment was out of order, but the command was sent up by Moscow regardless and unchecked. Fundamentally though, of course, the roots of the problem were deeper, for it was a badly designed system.

An investigation was ordered. This became psychologically quite problematical, for over-severe penalization of those responsible would have a demoralizing effect on the now extremely nervous Phobos 2 team. IKI director Roald Sagdeev urged a postponement of any disciplinary decision and mischievously quoted the notorious

Phobos 1 launch

secret service chief Lavrentin Beria who once said 'Let's make them work for now. We can still shoot them all later.' Much later, Beria's logic reasserted itself and the Yevpatoria commander was duly dismissed.

The chance of mission success was now reduced by half, literally at the stroke of a key. Moreover, the loss of Phobos 1 meant that there was no basis for comparing and calibrating scientific in-flight and Mars arrival data.

PRECARIOUS ARRIVAL AT MARS

All was now dependent on Phobos 2. Its in-flight equipment was in good order. Phobos 2's solar low-energy detector (SLED) picked up the energy of a large solar

Phobos mission control

flare in October 1988. The solar energy detector was based on a design used for the *Giotto* Halley spacecraft, but it had to be redesigned for Phobos because Phobos used a different computer system. SLED consisted of two tube-like detectors. When they picked up solar energy particles, they counted them and rated them on a scale of six. SLED was then interrogated for the data by the Phobos computer every 230 sec.

After 200 days and a journey of 470m km, Phobos 2 entered Mars orbit on schedule at 15:55 on 27th January 1989, becoming the fourth Soviet probe to do so. The burn lasted 200 sec and the high elliptical orbit measured 800 × 80,000 km with a period of three days. This time, news about its equipment was not as good. Even before orbital insertion, there had been reports of 'isolated malfunctions' in its systems. NASA had been following the mission through its deep space network and reported that transmissions appeared to have switched from high rate to low rate and that at least one of the Bulgarian cameras appeared to have failed. It seemed that the main transponder, which had 50 W power, had failed and a backup was being used with 5 W. Worryingly, the February meeting of the international scientific collaborative team was postponed until March and then deferred indefinitely.

Soviet/Russian spacecraft in Mars orbit

Spacecraft	Arrival date	Orbit
Mars 2	27 Nov 1971	1,380 × 24,938 km, 18 hr, 48.54°
Mars 3	2 Dec 1971	1,500 × 190,000 km, 12 days 19 hr, 60°
Mars 5	12 Feb 1974	1,760 × 32,560 km, 24 hr 52 min, 35°
Phobos 2	29 Jan 1989	9,560 × 9,760 km, period 8 hr, 0° (final orbit)

NASA's interpretation was accurate enough, even though the reasons were not known at the time. There were indeed computer malfunctions and problems with the high-speed transmitter, which between them reduced dataflow badly. The craft's onboard control system was based on three computer processors, using a system whereby two processors could outvote the third one, emulating a system successfully developed by the Americans for their Voyager spacecraft. By the new year, one of Phobos' processors was gone, the second was giving trouble and there was no provision for the third to operate on its own, a design error. The surviving computer was programmed to automatically presume that the other two were voting against it but it could not vote against these two 'dead souls'.

MANOEUVRING TO THE MOON

Achieving the main part of the mission was now a race against time and the degrading computer and transmission systems. Despite the earlier difficulties, Phobos 2 now began to make good progress.

In its high orbit, Phobos 2 made its first close approach to the planet Mars at 864 km on 1st February. The scientific equipment was turned on and data from the first periaxis were transmitted the following day.

Now began a series of complex manoeuvres to bring Phobos 2 to its target by early April. On 12th February, Phobos 2 raised its periaxis from 800 km to the height of the observation orbit and was now circling Mars at $6,400 \times 81,200$ km, 86.5 hr, $0.9°$. There was a worrying development when radio signals failed on 14th February, but they were quickly recovered.

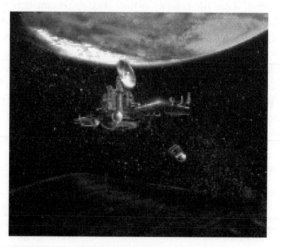

Phobos in Mars orbit

The second manoeuvre took place a week later on 18th February and was a critical one, placing Phobos 2 in its observation orbit. This reduced the apoaxis by a large margin and altered the inclination to make it more equatorial. The lower stage of the combined propulsion and braking system, the autonomous propulsion system (APS), now placed Phobos 2 in an almost circular orbit at 9,760 × 9,690 km, period 8 hr, 0.5°. With this main manoeuvre completed, the APS detached, its work done.

The observation orbit brought Phobos 2 close enough to take its first pictures of the moon Phobos on 21st February at a distance of between 860 km and 1,130 km. On 27th February, Phobos 2 aligned its position toward the moon, using the planet Jupiter as its navigation reference point. The idea was that Phobos 2 would remain on the lit-up side of the moon. The probe now came to within 320 km of Phobos and took 15 pictures. On 1st March, the TERMOSCAN device was first used to scan the surface of equatorial Mars. For the following week, the probe turned on its instruments for X-rays, scans of the solar disk, cosmic rays, gamma bursts and the Martian magnetic field.

On the 3rd March, scientists announced that they had now determined the orbit of the moon Phobos to an accuracy of 5 km. On 7th March, the CBPS second-stage engines were fired to equalize the orbital plane of the probe with the moon itself (0°). On 15th March, the engines were fired to adjust the observation orbit to one 200 km above that of the moon, but with two intersecting points in each orbit. This was refined in a third manoeuvre on 21st March.

Moon Phobos in sight

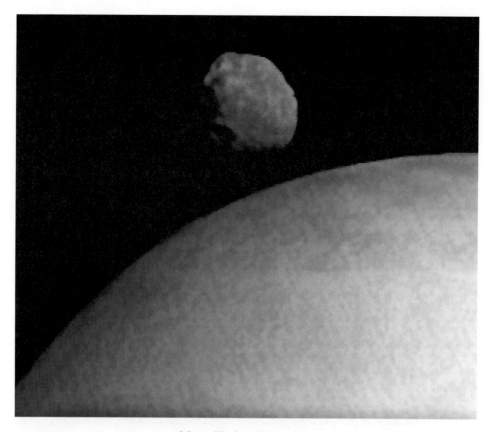

Moon Phobos closer view

Close approaches to Phobos
23 Feb   860 km
28 Feb   320 km
25 Mar   191 km

The full instrumentation suite again took observations on 22nd and 23rd March. At a distance of between 191 km and 279 km, visual observations were made of the moon over 80 min on 25th March to determine the best landing spot for the lander and hopper. TERMOSCAN was used on 26th March and further pictures were taken of the moon. The first communications session on 27th March lasted from 10:25 to 12:47 and a further round of pictures were taken of the moon, some as close as 400 km. The landing manoeuvre was being prepared, with 9th April set as the date for touchdown. By this stage, 40 images of Phobos had been obtained from distances of 200 to 1,100 km, covering more than 80% of the surface with a resolution of down to 40 m. The probe had now been 57 days orbiting Mars. Six new craters and eleven new grooves were found.

Jupiter seen by Phobos 2 in Mars orbit

Phobos 2 in Mars orbit: the manoeuvres
Autonomous propulsion system/lower-stage combined braking and propulsion system

27 Jan	Mars orbit insertion	$800 \times 80{,}000$ km, 72 hr, $1°$
12 Feb	Raise periaxis for observation orbit	$6{,}400 \times 81{,}200$ km, 86.5 hr, $0.9°$
18 Feb	Lower apoaxis for observation orbit	$9{,}760 \times 9{,}690$ km, 8 hr, $0.5°$

Upper-stage combined braking and propulsion system

7 Mar	Equalize orbital plane with moon	$9{,}760 \times 9{,}690$ km, 8 hr, $0°$
15 Mar	Establish two intersection points	$9{,}560 \times 9{,}760$ km, 8 hr, $0°$
21 Mar	Final refinement for intersection	$9{,}560 \times 9{,}760$ km, 8 hr, $0°$ (final orbit)

'THE LAST MESSAGE FROM THE DYING PHOBOS 2'

The next session was scheduled for 14:59 later that day for the imaging of Phobos at a distance from 214 km to 371 km. While imaging Phobos, the craft would be turned

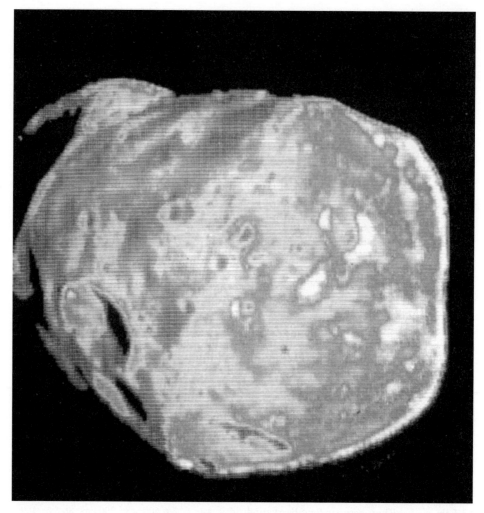

Moon Phobos close-up image

away from ground control and would then swivel around afterwards to relay the pictures back. Everything went fine.

The probe was then commanded to take further pictures between 17:59 and 18:05 and signal them back immediately afterwards (there was a 10 min delay due to distance). This time, once the session was over, signals were not reacquired. Ground control then sent up a set of commands to restore communications. Efforts were made for four hours. Signals were picked up again at 19:50 but for only 13 min. This was thought to be the omni-directional antenna sweeping slowly by Earth, not the high-gain antenna. As IKI director Roald Sagdeev was to lament later: 'There was a very weak, unintelligible signal indicating that the spacecraft was in an uncontrolled state. It was the last message from the dying Phobos 2.'

Phobos 2 last image of Moon Phobos

Twelve groups of engineers and scientists were at once convened to examine what could be done. The picture the controllers had was that Phobos 2 had lost its bearings and was rotating. Without solar lock, the batteries would discharge in five hours. The batteries drained and on 16th April the probe was formally declared abandoned.

The Phobos failures were a great shock to the interplanetary side of the Soviet space community. Not a single space probe had failed since the disappointments of the great Mars fleet and there had been an unbroken run of successes, culminating in

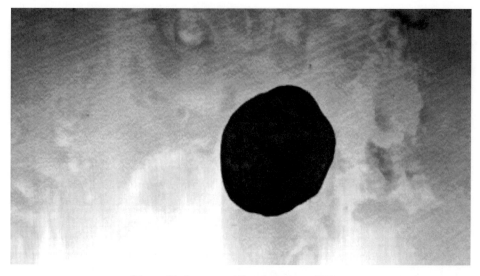

Moon Phobos over *Tharsis* region of Mars

VEGA. There was a great sense of loss, disappointment, emptiness made worse by the circumstances in which the missions had failed, one so close to its objective.

The subsequent commission of inquiry, established 31st March, set off an acrimonious round of recrimination within the Soviet scientific community, *glasnost* making possible the letting off of years of stored-up feelings and resentments. Scientists blamed engineers and engineers blamed scientists. IKI director Roald Sagdeev railed against the haste in which the project had been prepared, 3.5 years, half the time of the more successful VEGA. When the VEGA 2 computer guidance system had broken down a half-hour from interception, it switched quickly over to a backup, but the Phobos computer system had no equivalent system. This time, his comments and those of others were reported on openly in the increasingly liberated end-of-Soviet-period media. IKI pressed for an international investigation as the only way to restore credibility to the programme. *Perestroika* or not, the old guard fought back, reverting to its old ways of concealment, pleading that 'for the good of the programme' awkward issues be avoided.

When a dejected Phobos international team arrived in Moscow in May 1989 for the end-of-mission summary and *post-mortem*, Lavochkin deputy director Roald Kremev gave them explanations about the 'dangerous environment' around Phobos, such as solar flares, dust particles and meteorites. No one believed him, especially his own colleagues, never mind the international community. The immediate cause of the failure was that the probe had lost its bearings, the solar panels had stopped supplying power and the batteries had drained after five hours.

The Soviet system found great difficulty in admitting that there were deeper causes. As if to concede that there were some, Lavochkin director Vyacheslav Kovtunenko added that for future missions there would be better control and orientation systems, modifications of the computers and greater power in the storage batteries. There had been no fail-safe mechanism and no system to realign either craft with its star trackers after a certain period or to stabilize a disorientation.

His colleagues elaborated. Lavochkin deputy director Vladimir Perminov added that spacecraft designs before Phobos had a survivability factor built in called 'e-minimal'. When battery reached a certain level, a command would have been triggered to conserve supplies and protect a minimum level of systems. But that system was not fitted to this design. Sagdeev, for his part, criticized the ministry for not having an intermediary, supervising scientific team, like the one used on the VEGA mission, for this might have spotted the weaknesses. He also admitted that with such a long run of successes, some of the programme leaders may have become overconfident and taken their eye off the ball.

The main lessons learned were as follows:

- With Phobos 1, the computer software should have been scanned for errors, so as to make sure that fatal commands could not be issued by a single error. The message should have been checked before it was sent. The onboard computer should have been empowered to reject the signal as illogical at that stage of the mission.
- The surviving computer on Phobos 2 should have been empowered to vote against 'dead souls'.
- The need to have scientists and contractors involved together at a series of levels in mission planning and design.
- The importance of making spacecraft 'survivable'.

SCIENCE RESULTS FROM PHOBOS

Despite the heartbreaking outcome, so close to the landing on Phobos, there was a substantial scientific return [6]. This was evident the moment Phobos 2 entered Mars orbit, for instruments detected a bow wave 1,000 km above Mars, indicating a weak magnetic field. Interpretation of the data enabled characterization of the field, with four features: magnetopause boundary, shock wave, plasma layer and tail. The data from Phobos 2 appeared to indicate that Mars's magnetic field echoed that of Venus [7]. Although the planet might have a magnetic field, it was very weak, some 1/10,000th that of Earth. Instead, the ionosphere warped the passing solar wind into a bow shock, creating a magnetic tail behind the planet.

But, it was very weak and this weakness may have contributed to the planet losing its water. Phobos instruments indicated that Mars was losing its atmosphere at the rate of 2–5 kg/sec, which was significant granted its low density (on Earth, such an amount would be negligible). Water vapour content in the atmosphere of Mars between 20 km and 60 km was calculated at 1/10,000th that of Earth and the main component carbon dioxide. The atmosphere contained only 0.005% water. As for the surface, small but variable signs of water were indicated around volcanoes and on the surface around the *Vallis Marineris*. Phobos 2 confirmed indications from Mars 5 of

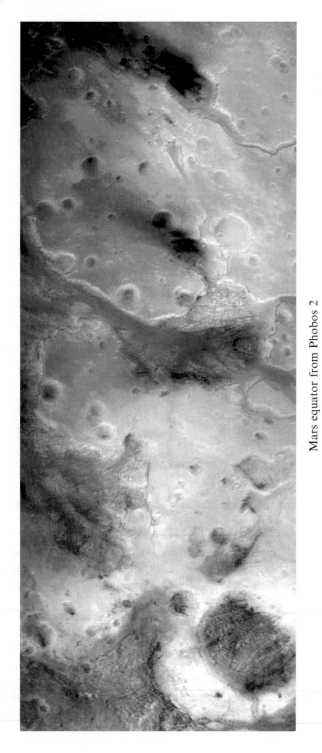

Mars equator from Phobos 2

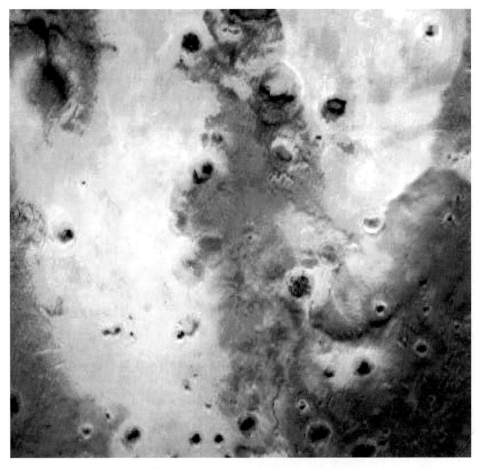

Phobos 2 image of chaotic Martian terrain

ozone in the atmosphere. The infrared spectrometer recorded the signatures of 33 sunrises and sunsets, detecting ozone at an altitude of 12 km.

The TERMOSCAN instrument made four passes over Mars, with a maximum resolution of 300 m and made the first thermal chart of Mars' surface. TERMOSCAN imaged the equatorial zone of Mars in a strip 1,500 km wide with a resolution of 2 km from an altitude of 6,000 km. TERMOSCAN measured surface temperatures ranging from $-93°C$ to $7°C$, with areas of local variation of $22-30°C$. Analysis of the cooling and reheating of the planet's surface as Phobos' shadow crossed Mars provided information on the thermal quality of the planet's surface, finding good insulating material down to 50 μm, but poorer below.

Four panoramas were taken: on 11th February, 1st March and 26th March (daytime and evening). Some show a small smudge – Phobos traversing the rusty Martian landscape below. Major surface features were identified and the height of some of the volcanoes was extrapolated (e.g., *Pavonis Mons*, 5.9 km). The more detailed pictures were taken by the VSK video-spectrometric complex built in Bulgaria, the GDR and Soviet Union, and these took, remarkably, a picture of distant Jupiter, the first time it had been seen from Mars orbit.

Two instruments contributed to an improved knowledge of the composition of Martian rock. The French infrared spectrometer suggested that it had a sedimentary nature. The gamma ray spectrometer, similar to one carried on Mars 5, was activated four times during close approaches of the transfer orbit at a distance of 3,000 km from the surface, at all times over the equatorial regions, where it made eleven tracks between 30°N and 30°S, covering such features as the *Vallis Marineris* and the largest volcanoes. As a result, it was possible to calculate surface composition.

Phobos 2: composition of Mars rock

Oxygen	48%
Magnesium	6%
Aluminium	5%
Silicon	19%
Potassium	0.3%
Calcium	6%
Titanium	1%
Iron	9%
Uranium	0.5×10^{-4}
Thorium	1.9×10^{-4}

Source: Surkov (1997) [8]

Broadly, these figures confirmed those of Mars 5. Combined with American Viking results, they also indicated that much of the area concerned was covered with a layer of fine-grained, wind-blown material and that these composition figures represent both the surface and the underlying bedrock [8].

The probe did not come close enough to the moon Phobos to reach many definite conclusions, but its daytime temperature was calculated at 27°C. The cameras took 37 pictures of Phobos itself, covering 80% of its surface, finding six new craters and eleven hollows not previously seen. The moon appeared to be made of carbonaceous chondrite, meaning that it was most likely a captured asteroid. Scanning of Phobos itself revealed a variety of surfaces, suggesting that the moon was a conglomeration of many different materials. The mass of the moon was estimated at 1.08×10^{17} kg and its density low, at 1.95 g/cm^3, suggesting either that it is made of porous material or had lighter material deep within (e.g., water ice) (or a combination thereof). The infrared spectrometer indicated the former, for it found no sign of hydration.

The Irish experiment, the solar low-energy detector (SLED), confirmed definable but low levels of radiation around Mars, below the threshold that would be dangerous to cosmonauts. Phobos 1 and 2 registered over a hundred bursts of X-ray and gamma-ray emissions, one of which, on 24th October 1988, was the most intense ever noted.

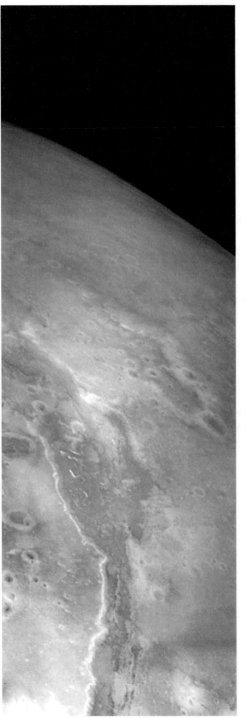

Phobos 2 over *Vallis Marineris*

Science from Phobos 2
Bow wave 1,000 km above Mars, weak magnetic field.
Four features of magnetic field: magnetopause boundary, shock wave, plasma layer, tail.
Loss of planetary atmosphere at rate of 2–5 kg/sec.
Water vapour content 1/10,000th that of Earth.
Levels of radiation safe for cosmonauts.
Temperature of Phobos surface, 27°C,
Mapping of surface of Phobos.
Composition: carbonaceous chondrite.
Composition of Martian surface rock.
Thin layer of blown sandy material covering Martian surface.

PUTTING IT BACK TOGETHER AGAIN

The failure of Phobos was a debating point during that spring's elections to the Soviet parliament. Long-time space commentator Boris Belitsky noted: 'The failure of Phobos 2 damaged space research in the eyes of the public. In the general election, several candidates proposed cuts. Failure came badly to people fed on a diet of success.' In April 1990, in response, the government cut the Soviet space budget from 300m roubles a year to 220m.

There was a lengthy internal debate about whether to try a third mission to Phobos or whether it was better to return to the Martian surface. Several in the Institute of Space Research, IKI, still hoped to send the ready-to-go backup model, Phobos 3, to Mars in 1992, but this proposal did not win out. Ultimately, the probe ended up going for auction in the West to raise hard currency, even though it was an operational rather than an engineering model. Years later, several scientists took the view that the wrong decision was made here [9].

The post-Phobos period saw many permutations in plans for the Soviet Mars programme, complicated by different priorities and desires in the Institute for Space Research, the Vernadsky Institute and other institutional players. Even before Phobos 1 and 2 had taken off, follow-on missions had been discussed. The first was mooted in 1987, when the Phobos mission was at an advanced state of preparation. The original idea was 'Project Columbus', using two Protons to send two spaceships to drop balloons and landers with rovers from Mars orbit in 1992, on the 500th anniversary of the visit by Christopher Columbus to the new world in 1492. *Columbus* would conduct a full survey of the planet for a year while rovers drilled for rocks down below [10].

Nothing was going to happen until the government approved the mission and allocated money. This did not happen until late 1989, when the government gave the go-ahead for a mission in 1994, meaning that the 1992 window would be missed. Now, presumably too late to mark that anniversary, the term *Columbus* was dropped from the interplanetary planning lexicon. What was now called Mars 94 would be followed by an orbiter and rover in 1996 and a sample return in 1998–2001.

In 1990, new schedules were published. These were now for two polar orbiters mapping the surface and deploying a balloon in 1994, rovers in 1996 and a soil sample return mission in 1998. Much of the public attention focused on the planned French balloon, which would float up to 4,000 m high in the sky in the day but descend to the surface by night. For the 1996 missions, two rovers were under consideration: one of 250 kg, one of 600 kg, able with radio-isotopes to travel for three years for distances over 500 km.

The balloon mission was a creative project. Standing 50 m tall, the balloon would hold 5,000 m^3 of helium gas. With a diameter of 13.2 m, its overall payload weight was 258 kg. The helium was in an ultrathin 6-μm envelope. Underneath was a gondola and below that a thick, slithering rope containing instruments, called a 'snake'. The 20 kg gondola would be equipped with a TV system, altimeter, magnetometer and infrared spectrometer, while the snake would have 3.4 kg of instrumentation, comprising gamma spectrometer, pulsed radar accelerometer and dosimeter. These instruments would measure temperature, pressure, humidity, dust levels, wind and take pictures of the panoramas passing below. The balloon would inflate 11.4 km over the planet's mid-northern latitudes, descend to an operating daytime altitude of between 2 km and 4 km, landing on the surface by night. It was expected to achieve 10–15 landing and take-off cycles and cover several thousand kilometres. Tests of a model of the balloon were carried out by Soviet, French and American engineers in the Mojave desert [11].

The first funding for Mars 94, 20m roubles, arrived in April 1990 as the initial instalment of its estimated 500m rouble budget which was intended to cover two spacecraft, including rovers and penetrators. A further 13m roubles followed in July. International participation had already been formally invited at a seminar held in Moscow in November 1989 with guests from Austria, Bulgaria, Finland, France, Germany, Britain, Ireland, Sweden and the United States. The key countries of Germany and France quickly joined, agreeing to contribute $120m in equipment.

At this stage, the broader financial problems affecting the Soviet Union began to have a visible effect on the programme. The fact that people were noting the arrival of

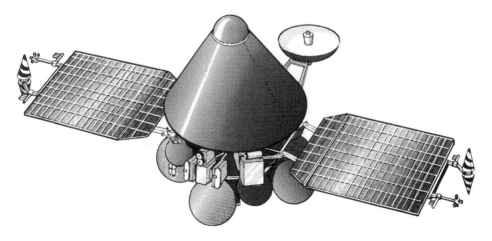

Mars 94

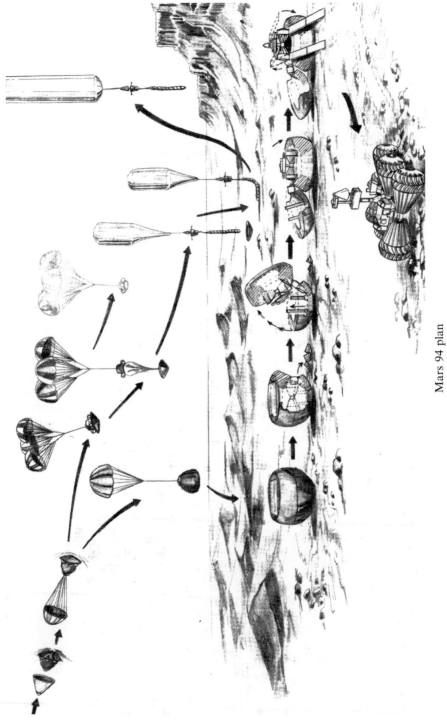

Mars 94 plan

designated funds was a warning sign in itself, for until then the money had always just arrived as a matter of course. A year later, in April 1991, the two planned spacecraft were divided into one for the 1994 window, the second for the 1996 window. Mars 94 would be a single spacecraft with surface stations and penetrators, while Mars 96 would now carry a rover and the now-delayed balloon.

THE SOVIET UNION BECOMES RUSSIA

The great 1924 film *Aelita* had seen a revolution establishing Soviet power on Mars. Back on Earth, another revolution disestablished Soviet power, which ended peacefully on new year's eve at the end of 1991. With it went the Ministry of Medium Machine Building. Sergei Afanasayev had ruled there from the day of its founding in March 1965 until he was moved involuntarily in a ministerial shuffle in April 1983. The place of 'Big Hammer' was taken by the more mannerly Oleg Baklanov, who later joined the group of conspirators who organized the *coup* of August 1991. He ended up in prison afterwards.

As the Soviet Union became Russia and the Commonwealth of Independent States on 1st January 1992, the problems multiplied. Progress with the Mars 94 project slowed and the instrument package was not finally agreed till early 1993. Assembly of the now single spacecraft continued on a stop–go basis, stopping when the money fell short or parts were not delivered, especially from the former Soviet republics whence they now had to be imported. Russia appealed for Western help to keep the programme going. Germany and France now paid in an additional $10m to protect their investments and this prompted a fresh surge in assembly in late 1993. Attempts to involve the Americans, despite the endorsements of Presidents Bush and Yeltsin in summer 1992, came to nought and the Americans decided that as far as Mars 94 was concerned, they would not come out to play.

An early decision of the new government was to establish a Russian space agency, the RKA. It was soon confronted with the difficult choice of what to do with Mars 94: to deliver a badly prepared spacecraft on time, or to delay a further two years, with the risk that it might never be finished. In May 1994, they chose the latter, so Mars 94 now became Mars 96. On the proposal of Vyacheslav Kovtunenko, the new space agency descoped the project, with the balloons removed. As for the original Mars 96, this now became Mars 98. The upper stage would be a lander in a large, elongated cone, with the lower stage staying behind in 350 km polar orbit and a 350 kg experimental package. It would land using a mixture of parachute and engines.

Possibly to make the Mars 96 decision seem more convincing, a date was even set: 16th November 1996. The Russian government declared all along that Mars 96 was a top-priority, showcase project. According to analysts Burnham and Salmon, had it not been for the high level of international participation and the associated legal obligations and cash involved, the project would most likely have been cancelled altogether [12]. It is hard to disagree.

Even within the bigger picture, there were mini-crises. The Russians tried to save expense by replacing the new Argus mobile scan platform with a fixed system installed on the main body of the spacecraft, as was the case in the days prior to VEGA. The Germans, who were responsible for the cameras, were furious and reckoned that this would sharply reduce their effectiveness. The problems were only solved when the Germans sent a team of technicians to St Petersburg Institute of Precision Mechanics and Optics to resolve the scan platform problem.

In the midst of all this, Lavochkin chief designer Vyacheslav Kovtunenko died on 11th July 1993, aged 73, after 15 years at the helm. He was replaced by bureau veteran Stanislav Kulikov.

COMPLETED BY CANDLELIGHT

Even with the deferment, keeping the project going was a major challenge. Only $122m of government money ever arrived for the project, all of it late and most taken from other projects, such as the Spektr Observatory and the now-cancelled Luna 92 moon probe. Foreign countries actually contributed much more by the very end – about $180m. The project was repeatedly in an on/off situation with threats of further delay or even cancellation.

The financial situation deteriorated to the point that money was no longer there to support the ocean-going tracking fleet. Its ships were recalled to port in February 1992. The *Borovichi, Morzhovets, Kegostrov* and *Nevel* were sold while the *Cosmonaut Vladimir Komarov* was turned into an ecological museum. The *Cosmonaut Yuri Gagarin* and *Academician Sergei Korolev*, whose home port was Odessa, became part of the Ukrainian naval defence forces who tried, unsuccessfully, to sell them in 1994. Mission planners still hoped to send one tracking ship, the *Cosmonaut Viktor Patsayev* to the Gulf of Guinea to follow Mars 96 at the crucial point of trans-Mars insertion.

The Mars 96 science platform was eventually assembled in late 1995 and integration tests began in Lavochkin in January 1996, concluding in June 1996, despite the bureau being $80m out of pocket from delayed government payments. The spacecraft was shipped to Baikonour during the summer. Even then, finishing the probe off in the hangar in Baikonour presented its own problems. Baikonour was suffering from power shortages as utilities tried to keep going in the face of unpaid bills. At one stage, in a telling parable of the state of Russia at the time, there was no electricity, so technicians found themselves trying to complete the ultramodern Mars probe in the cold and the dark by candlelight and kerosene heaters. Many engineers were never paid for their work.

Meantime, the broader picture continued to worsen. In early 1995, the government decided that a Proton could not be made available for Mars 98, for such a launcher was now too expensive. Mars 98 was downscaled to use the old 8K78M launcher, retired from the interplanetary programme in 1972. This reduced the size of the payload dramatically, from six to 1.2 tonnes. The two payloads were reduced proportionately, the French balloon to 258 kg and the rover to only 100 kg. Even so, Mars 98 was cancelled later in the year. For many years, there had been talk of a Phobos sample return, a network of small stations on the planet (Mars Glob) and planning for some of these later missions actually reached quite an advanced stage

Mars 96 in assembly

Mars 98 original plan

[13]. Hopes for such missions now receded into the far distance. By the time Mars 96 made it down to the pad, it was recognized that this would be the last Russian mission to Mars for a very, very long time.

AFTER MARS 96?

At this stage, as Mars 96 was finally readied for take-off, it is worth looking ahead to see what the Mars 98 mission was to have achieved. The Russians were keen to build a rover to explore the Martian surface. Two were designed, each with six conical wheels. The first was for a larger rover, between 350 kg and 500 kg in mass, with a scientific payload of up to 70 kg and a speed of up to 2 km/hr. Wheel diameter was 50 cm and speed 2 km/hr. Base was 2.9 m, clearance 360 mm.

In the event, VNII Transmash was told that a mission to land a rover alone was not realistic: a rover would have to be part of a multiple payload. VNII Transmash was told to reduce the ambition of the rover from 500 kg to around 100 kg, with a launch date pushed back first to 1996 and then 1998 [14]. Then, Mars 98 would see the landing of a 100 kg Mars rover (Marsokhod) in the *Arcadia* region, powered by a radio-isotope nuclear generator to explore 100 km distant areas on a year-long mission, carrying eight scientific instruments of 15 kg each. Wheel diameter was 35 cm, speed 500 m/hr. The Marsokhod would be ideal for getting to hard-to-reach locations.

Tests were made of this Marsokhod in 1993 not only in Russia but also in Toulouse, France and in the Mojave desert and Death Valley in California by McDonnell-Douglas with colleagues from Hungary and the United States. The western American desert had the advantage of having dunes, sand and river beds, features that such a vehicle would encounter on Mars. The full-scale model had six wheels, 35 cm in diameter (the intended size), designed to negotiate sand and awkward

Mars 98

surfaces, a high platform and solar panels at the rear. French mobile navigation experts helped to develop an autonomous navigation system. The rover was designed for two possible profiles. With a conventional lithium battery, it could travel 30 km over 100 days at up to 500 m an hour. Alternatively, with a 6 W radio-isotope, it could travel 180 km for a year. It would carry a penetrator able to go 5 cm down into the soil and equipment to sample rocks [15]. The use of conical wheels was based on the Lunokhod experience: such wheels were more versatile, were better climbers and could enable the vehicle to bridge a crevasse.

After the conclusion of tests, the Marsokhod was exhibited at a press conference in Los Angeles. At one stage, the rover fell into a large trailer tyre, but to the cheering of journalists, managed to diagnose the situation and climb out [16]. During 1991 – in fact, during the period of the *coup* in the Soviet Union – the Marsokhod was jointly evaluated by American scientists through the Planetary Society and Russian experts from NPO Lavochkin, the Institute of Volcanology, IKI and VNII Transmash [17]. Although more remote than the Crimea, where the lunar rover Lunokhod was tested, Kamchatka in the far east was the most volcanic landscape in the Soviet Union and had many features similar to the imagined Martian landscape. Reachable by helicopter only at the end of the summer, Transmash had a camp of wooden buildings high in its bleak landscape. During 1991, the Mars rover was tested for manoeuvrability under the control of a human operator in preparation for the installation of computer controls.

Mars 98 Marsokhod

Consideration was given to the idea of Marsokhod carrying a micro-rover, weighing in the order of 4–10kg. The idea was that a small micro-rover could reach sampling points unavailable to the larger rover, test out problematical sites in advance of the larger rover and explore around the large rover if it were stuck so as to determine the best route out [18]. The micro-rover would have its own science package and could travel up to 80 m a day up to a grand total of 10 km.

Mars 98 would also drop the long-planned French balloon. Now, its height had declined to 42 m, and smaller versions 21 m high were also under consideration. Another approach was to put the instruments in the snake with a soil analyzer – not in a gondola.

Mars 98 was to have been the final step before a soil sample mission during the next Mars window. Again, paper studies were carried out on this, building on the experience of the 5M and 5NM projects but using the UMVL design (which, lacking the Proton, became ever more theoretical). The designers agreed that, because the engine of the landing craft would disturb the surface of the landing area, samples should be collected some distance from the landing craft. The use of roving vehicles was integral to the sample return missions. The preferred approach was that a rover would collect samples over an extended period and then act as a beacon to guide in the rocket that would bring these samples back to Earth. An alternative approach was to make the return rocket an integral part of the rover. Once enough samples were collected and transferred, then the small return rocket would take off from the rover and head back to Earth. But this would need a large vehicle, weighing something in the order of a tonne. Options were kept open on three issues:

- whether the return rocket should be in the landing craft, with the rover to deliver samples to the landing craft, or on the rover itself;
- whether the return rocket should fire directly back to Earth, or whether it should rendezvous with a second orbiter in Mars orbit which would in turn bring it back to Earth (the second was favoured).
- whether the return craft should reenter the atmosphere directly, or enter Earth orbit for subsequent rendezvous, docking and decontamination (the second was favoured).

After a 350-day flight to Mars orbit, the lander would descend, like Mars 98, carrying either the return rocket and rover, or the rover with the return rocket built in.

MARS 96: LAST OF THE HEAVYWEIGHTS

Meantime, Mars 96 had at last reached its winter launchpad. This was a truly gigantic spacecraft, 6,640 kg, including a payload of 645 kg and three tonnes of fuel. It was an ambitious mission, arguably one of the most adventurous planetary missions ever mounted by any nation, matching the American/European probe *Cassini* to Saturn launched the following year. No fewer than 20 countries eventually participated in the experimental packages.

Mars sample return

The purpose of the Mars 96 mission was to place the orbiter over Mars in September 1997, equipped with cameras, sensors, relay systems and 24 experiments; to send down two small landers on Mars; and to drop two penetrators each carrying a camera and miniaturized sensors which would tunnel into the surface. As evidence that some lessons had been learned from the Phobos experience, more advanced computers were supplied, but from the French and European space agencies who did not want to take any chances. The solar arrays were made larger.

Mars 96 was to take a long, slow, curving trajectory to the planet. Because the station was so heavy, the combined braking and propulsion system (CBPS), already well proven on the Phobos mission, would be required to give the station a final kick on its way out of Earth orbit for trans-Mars injection and would fire again for a big burn at Mars orbit injection in September 1997. As it travelled to Mars, a suite of instruments would study radiation, solar activity, gamma bursts and the stars. The objective was to enter a polar orbit covering the whole planet (300 × 22,577 km, 106°, 43 hr) and then burn again to adjust the orbit to working altitude of 250 × 18,000 km, 101°, which it would reach in May 1998 and last until 2000. This would come as near as 200 km at its closest point.

Thirty instruments were carried on the nuclear-powered orbiter, designed to make weather studies and scan below the surface for water. The most prominent was the Argus television complex, so named after the many-eyed giant of mythology. Argus, made by the Institute of Precision Mechanics and Optics in St Petersburg, would scan the Martian surface both from altitude with wide-angled lenses and from as close as 200 km with zoom instruments in 500 km swathes. Argus's resolution was 10 m, and the purpose was to compile a complete mineralogical and topographical map of the planet. At the same time, a videospectrometer would compile an infrared map of the surface. Another TERMOSCAN would make a temperature map of the Martian surface. The orbiter carried an altitude radar to map the mean level of Mars and eight spectrometers to measure everything from cosmic rays to ions, plasma, neutrons and particles.

Two 65 kg penetrators – each 1.5 m long, 120 mm in diameter – would be released after a month in orbit, spun up to 95 rpm, enter the atmosphere at 4.9 km/sec at an angle of 10° to 14° and crash-land in *Arcadia* and *Utopia*. Inflated gas bags would break their fall, but the penetrators were still expected to impact at some speed, stopping at up to 80 m/sec, or up to 1,000 times the force of gravity. Many tests were run to verify the landing technique, including dropping the package 60 m down the liftshaft of the Moscow Aviation Institute and later from helicopters. Out of the tail would peep a camera, magnetometer, transmitter, weather detectors and wind meter. The front end of the penetrator would dive into the Martian soil as deep as 6 m below the permafrost layer. The underground part was equipped with an accelerometer, X-ray spectrometer, gamma-ray spectrometer, alpha-particle spectrometer, thermometer and neutron detector, which would relay their findings to the transmitter about the surface through a cable rope. The penetrators would transmit for a year.

The most exciting part of the mission, though, was the release of the Mars landers. The two small 33 kg landers, shaped like Luna 9 and 60 cm in diameter, would separate, enter the atmosphere at 5.75 km/sec at an angle of between 10.5° and 20.5°, parachute down from 19 km and, encased in inflated airbags, bounce over the surface as high as 70 m before settling. Once stable, the spacecraft would open up and begin transmitting photographs, to be relayed back by the orbiter on its daily passes. On each lander was a compact disk called *Visions of Mars*, a time capsule containing the sum of Earthly knowledge about Mars at that point, including the famous broadcast of H.G. Wells's *War of the worlds* in 1938 which, when transmitted on

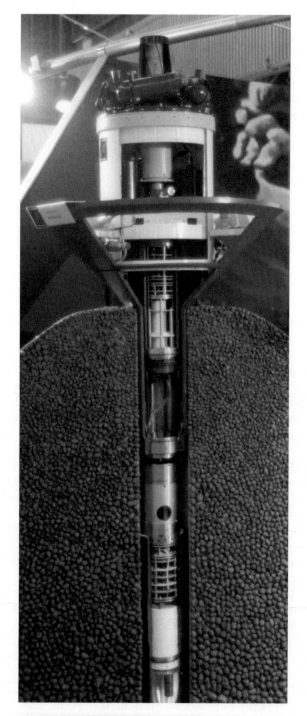

Mars 96 penetrator

American radio at the time, caused panic along the east coast. Each small lander carried a weather meter (Finland), thermometer, seismometer, landing and panoramic camera (Russia) and an alpha proton X-ray spectrometer (Germany). They would transmit data back to Earth for a Martian year, almost 700 Earth days, using the orbiter overhead. Their landing sites were:

41.31°N, 153.77°W	*Arcadia*
32.48°N, 169.32°W	*Amazonis*
3.65°N, 193°W	(reserve)

The landers would communicate with the orbiter daily in 10 min, 32 Mbit bursts at the beginning and then for 20 min for three days. The two penetrators and two landers would form a four-point seismic network.

En route to Mars, Mars 96 carried gamma ray and ultraviolet spectrometers and instruments for solar, plasma physics and stellar observations. A photometer called ERVIS was designed to focus on between 10 and 20 different stars and measure them for brightness and oscillations. Two radiation detectors were carried, one American (which had already flown on the shuttle) and another French (which had already flown on Mir).

Mars 96 surface station

Mars 96 instruments: in flight
Photon gamma ray spectrometer.
Neutron spectrometer.
Quadrupole mass spectrometer.
Energy mass ion spectrograph and neutral particle imager.
Omni-directional ionospheric energy mass spectrometer.
Ionospheric plasma spectrometer.
Electron analyzer and magnetometer.
Plasma wave detector.
Low-energy charged particle spectrometer.
Cosmic and solar gamma burst spectrometer.
Solar spectrometer.
Stellar spectrometer.
Stellar oscillation photometer.
Radiation and dosimetry control.

Mars 96 instruments: orbiter
High-resolution scanning camera (Russia, Germany).
Wide-angle optical scanning system (Russia, Germany).
Planetary Fourier infrared spectrometer (Italy, Russia, Poland, Germany, Spain).
TERMOSCAN mapping radiometer (Russia).
SVET high-resolution mapping spectrophotometer (Russia, USA).
SPICAM optical multichannel spectrometer (France, Russia, Belgium).
Ultraviolet spectrophotometer (Russia, Germany).
Long-wave radar (Russia, Germany, USA, Austria).
PHOTON gamma spectrometer (Russia, USA).
Neutron spectrometer (Russia, Romania).
Mass spectrometer (Russia, Finland).
ASPERA mass spectrometer and particle imager (Russia, Sweden, Finland, Poland, USA, Norway, Germany).
FONEMA ion and high-energy spectrometer (Russia, Britain, Czech Republic, Slovakia, France, Ireland).
DYMIO ion spectrometer (France, Russia, Germany, USA).
MARIPROB plasma and ion detector (Austria, Belgium, Bulgaria, Czech Republic, Slovakia, Hungary, Ireland, Russia, USA).
MAREMF electron spectrometer and magnetometer (Austria, Belgium, France, Germany, Hungary, Ireland, Russia, USA).

Mars 96 instruments: lander
METG meteorology instrument system – temperature, pressure, humidity (Finland, France, Russia).
DPI accelerometer, pressure, temperature (Russia).
APX alpha particle, proton and X-ray spectrometer (Germany, Russia, USA).
OPTIMISM seismometer, magnetometer (France, Russia)
Panoramic camera (Finland, France, Russia).
Descent camera (France, Russia).

Mars 96 instruments: penetrator
TV camera (Russia).
Meteorological system – temperature, pressure, wind speed and direction, humidity (Finland, Russia).
PEGAS gamma ray spectrometer (Russia, Germany).
ANGSTREM X-ray spectrometer (Russia, Germany).
Alpha photon spectrometer (Russia, Germany).
Grunt accelerometer (Russia, Germany).
TERMOZOND temperature probe (Russia).
KAMERTON seismometer (Russia, Britain).
IMAP magnetometer (Russia, Bulgaria).

INSTEAD OF MARS, THE ANDES: THE SHORT, SORRY FLIGHT OF MARS 96

Despite all the difficulties, the Russians eventually managed to launch Mars 96 on 16th November 1996. The Proton rocket soared eastward into the cold, clear night sky and the scientists at the launch site celebrated as what was now named Mars 8 entered its parking orbit of 51.5°, 139 × 155 km. But, come dawn, it was all over. Approaching Africa an hour later, the block D upper stage should have fired the spacecraft into a highly eccentric orbit, at the end of which the CBPS stage on Mars 8 would kick the spacecraft on its final route to Mars. Arrival at Mars was scheduled for 23rd September 1997.

It is unclear whether block D failed to ignite at all or burned for only a short period, possibly 20 sec and then shut down. It seems that Mars 8's payload, pre-programmed to believe that the trans-Mars manoeuvre had now taken place, separated from block D and used its engines to make what should have been the final kick burn to adjust its course toward Mars. The solar arrays were deployed. But, instead of kicking out of Earth orbit on its way to Mars, the CBPS on its own – without the benefit of block D – had given it a high point in Earth orbit of 1,500 km over Africa. It was an unstable orbit, with a low point of 87 km on the other side, over South America.

Once the trans-Mars manoeuvre was accomplished, Mars 8 radioed back to Yevpatoria the successful accomplishment of the burn. Maybe ground controllers thought for a moment that all was well, but it must have soon dawned on them that Mars 8 was nowhere on the road to Mars but spiralling back to Earth, as was the errant block D. Block D had a small beacon, but it is not known whether the Russians picked up its signal or not. The following days were a period of some confusion, made worse by the delapidated state of the Russian tracking network. Initially, based on ample years of experience of watching stranded Mars, Venus and moon probes, most observers reckoned there was only one object in Earth orbit, an unfired block D with Mars 8 still attached. Then the Russians let it be known that they had received separate signals from Mars 8 in a different orbit. Most observers agreed that if block D had fired, albeit briefly, then separation could have occurred. The only country with a worldwide tracking network was the United States, which had no information from the Russians about the two objects or which was which, the block D or Mars 8.

The Americans focused their efforts on tracking block D, which they still believed also contained an unseparated Mars 8. The Americans were of course concerned about the 200 g of plutonium on Mars 8 and where it would impact. They followed block D closely, convening the Emergency Response Team to prepare for the worst. At one stage it looked as if block D would impact on Australia. President Clinton telephoned Prime Minister John Howard to assure him of American assistance in search and recovery operations. In the end, block D stage crashed into the Pacific Ocean near Easter Island a day later and Australia was spared.

Block D was down, but the problem of course was that Mars 8, with the deadly plutonium on board, was still very much in its erratic orbit, which had a low point of 87 km, which was unsustainable for more than a few days. US Space Command had missed the separation of Mars 8 and never found it in orbit, having focused all its efforts on block D. A day after block D came down, the Russians stated that – although they knew that there had been a separation – they did not know where Mars 8 was either. Both countries did know that the low point of the orbit was over South America. The State Department in Washington DC told several South American governments that something might fall on top on them, but there were no buddy-to-buddy offers of assistance like the ones to John Howard. On the Russian side, Yuri Koptev told some South American ambassadors that – even if something did fall on top of them – there was no danger of contamination anyway and not to worry.

Mars 96's reentry was eventually spotted over southern Chile as a bright, slow-moving meteorite for almost a minute, with pieces of débris breaking away from time to time. As chances would have it, the skies were clear and it was now dark. Many calls came in to the government in Santiago from the rural areas, but it appears that they may have been dismissed as the imagination of unreliable and excitable rural peasants. But they were also seen by a number of professional observers, even a vacationing astronomer, who formally logged information on all the sightings. Between them, they were able to plot the ground path of the disintegrating Mars 8. It is believed to have crashed into the Andes mountains near the town of Iquique on the border with Bolivia only hours afterwards. Some of the equipment was designed to survive a tough Mars entry and landing, and there was speculation that the probes may have ended up in the Atacama desert, searching for life on Earth!

The Chileans were furious with the Russians for not warning them of its hazardous cargo, but – lacking tracking ships – the Russians found it difficult to predict the reentry area. Eventually, US Space Command admitted that Mars 8 probably did come down on land. The American reaction contrasted with the crash of another Soviet spaceship, the military reconnaissance satellite Cosmos 954, onto Canada in 1978. The Americans mobilized huge forces in the subsequent *Operation Morninglight* and successfully recovered most of the wreckage of Cosmos 954 and its nuclear reactor. But, Mars 8 contained no military secrets and Chile was so much farther away than neighbours in Canada. The Americans pointed out that Mars 8 was not their spacecraft or responsibility anyway, which was perfectly true. Russia was responsible, under international law, for search and clean-up, but their position is that they did not have the financial resources to do so. But, no attempt was made to warn the people of the Chilean or Bolivian mountains about how to identify fallen spacecraft, or what to do with plutonium remains. They are most likely still there [19].

Summary of planned and actual Soviet and Russian landing points on Mars

Probe	Latitude	Longitude	Region
Mars 2	45°S	213°W	*Eridania*
Mars 3	45°S	158°W	*Electris* and *Phaetonis*
Mars 6	24°S	19°W	*Erythraeum*
Mars 7	43°S	42°W	*Galle* and *Argyre Planitia*
Mars 96 first station	41°N	154°W	*Arcadia*
Mars 96 second station	32°N	169°W	*Amazonis*

The Mars 96 disaster came at the worst possible time for the Russian space programme, with finance and morale already at rock bottom. An atmosphere of deep gloom prevailed. An investigating commission reported in January 1997. The exact cause of the block D failure was never determined, although as many as 20 possible sequences of events were explored which might have explained it. The Russian Space Agency had been unable, in the end, to find the money to send the *Cosmonaut Viktor Patsayev*. This would have cost 15m roubles and that sum was not available. While the tracking ship might not have saved the mission, it would have been able to receive the block D telemetry during the crucial moment. The cause was put down as a random failure. The investigation does not appear to have been pursued with much determination, for the investigators knew in their hearts that there would be no more Mars probes to which the lessons arising from their investigation could be applied.

Some Mars 96 instruments did find their way to Mars in the end, for replicas later flew on other spacecraft. The high-energy neutron detector (HEND) was flown, successfully, on the American Mars Odyssey in 2001 and enabled maps to be compiled on the location and depth of water ice on the planet [20]. Other equipment was flown on the American Mars Polar Lander (which crashed) and Europe's Mars Express (which succeeded).

Mars 96 was not the only Mars probe that failed during the 1990s. Indeed, in a dark humour, some mission planners began to talk about a 'great galactic ghoul' that inhabited the Martian regions and snapped up passing spacecraft. The ghoul was well fed. The Americans attempted to return to Mars in 1993, their first probe there since the Vikings in the 1970s. Mars Observer was a sophisticated $1bn orbiter to scan Mars in detail from a number of paths over a full Martian year. Just as it was due to enter Martian orbit in August 1993, contact was suddenly lost and never regained. Nobody knew why, but the engine may have exploded due to a pressurization fault at the point of orbit insertion. Even more embarrassingly, the Americans lost a Mars orbiter in 1999 because ground control confused its trajectory measurements between metric and imperial units. Worse followed soon after, when the Mars Polar Lander fired its retrorockets to soft-land near the south pole. The subsequent *post-mortem* suggested that the jolt of deploying its landing legs gave a vibration similar to that of the landing, which wrongly triggered a signal to tell the computer that it had already touched down. This caused the retrorockets to turn off, leaving the craft to fall to destruction from a height. Ever since, subsequent orbiters have tried to spot the wreckage.

During the 1990s, the trend swung decisively away from large spacecraft of the type represented by Mars 96 and Mars Observer. In the United States, NASA Administrator Dan Goldin took a new approach. Alarmed by the fact that larger interplanetary spacecraft took so long to design and construct (up to ten years), their considerable cost (over $1bn) and that a single-point rocket failure could ruin the whole project (as Mars Observer and then Mars 96 amply demonstrated), he announced the new philosophy of 'faster, cheaper, better'. New planetary probes would be smaller, developed more quickly and launched more frequently, in the expectation that, even if there were some losses, more data would ultimately be returned at less overall cost and risk.

Goldin's approach came to inform American deep space exploration in the 1990s. The United States launched two spacecraft to Mars at the time of Mars 96: Pathfinder and Mars Global Surveyor, both on the smaller, cheaper and highly reliable Delta II rockets. Pathfinder came crashing through the Martian atmosphere on 4th July 1997, using a combination of cone, parachute and rocket, but encasing the lander in inflatable airbags of the type used by Luna 9 and 13 to reach the moon in 1966. Pathfinder bounced as many as 17 times before coming to rest, deploying a small rover called Sojourner. Mars Global Surveyor used a combination of rocket engine and aerobraking to enter an orbit where it could map the Martian surface, a task begun in 1999 and which lasted many years. Mars Global Surveyor was, in the course of time, joined by two more orbiters, Mars Odyssey in 2001 and Mars Reconnaissance Orbiter in 2006. The techniques used to land Pathfinder were repeated in January 2004 with the landings of the Spirit and Opportunity rovers. Given a 90-day liftetime, they crawled over the planet for over two years, climbing hills, venturing into craters, crossing sand dunes and traversing deserts, two of the most remarkable robots of their age.

HARD TIMES

Mars 96 marked the end of the Soviet planetary exploration programme for some time. Mars 96 was the 54th Soviet/Russian interplanetary spacecraft.

The Russian economy took a series of nosedives during the 1990s. The space programme contracted severely and entire fields of endeavour were abandoned. In reality, budget restraint had already begun during the final years of the Soviet period. The mid-1980s had probably represented a peak of employment in the industry, with 400,000 people engaged. Budgets were first reduced in 1990, after the parliamentary elections. In September 1991, the month after the *coup*, the space shuttle programme was delayed indefinitely, as was the Mir 2 space station. The lunar polar orbiter (Luna 92) and the star-mapping mission – the *Lomonosov* star mapper – were cancelled.

When the Soviet Union broke up in December 1991, an agreement was reached that space should continue to be a cooperation area for the new Commonwealth of Independent States. A new Russian Space Agency, the RKA was established. At first, neither move appeared to do much to halt the downward slide. By 1992, it was evident that the programme was in serious trouble. The proportion of the budget spent on space fell from 1.5% of gross national product at its peak to less than 0.23% now. Federal government allocations arrived late, if indeed they arrived at all. The new RKA space agency director and himself a former Lavochkin Mars programme engineer, Yuri Koptev, pleaded for resources, letting it be known that a third of military launchings were now behind schedule and a half of civil missions. The numbers employed fell to 300,000. The Energiya/Buran programme, the culmination of nearly 20 years of investment, was cancelled on 30th June 1993.

The financial crisis plumbed extraordinary depths. The memorabilia of the programme were sold in Sotheby's to make a few dollars, even Sergei Korolev's slide rule and Vasili Mishin's diary falling under the hammer (and Phobos 3, as noted earlier). Advertising was painted on the sides of rocket and cosmonauts did space-walks with inflatable Pepsi containers. Sergei Korolev's design bureau, the Energiya company, was privatized. The infrastructure of the programme declined, parts of the Baikonour cosmodrome becoming quite delapidated. The famous space shuttle Buran was destroyed when its neglected roof eventually collapsed on top of it. Even the manned programme found itself in serious trouble. Flights were delayed because nozzles or nosecones were not available while cosmonauts were kept orbiting, waiting for the money to arrive for the rockets to bring them home. By 1998, with the second great rouble crash, employment in the space industry was down by three-quarters, now to only 100,000. Workers were paid months in arrears and many enterprises were, on paper at least, bankrupt. Contractors went unpaid for long periods and Plesetsk cosmodrome was blacked out by exasperated utility companies when it could not pay its electricity bill.

The sharp contraction did not affect all parts of the programme evenly. Some parts were worse hit than others. The most severely affected parts were research and development, which almost ceased; the interplanetary programme; and some types of applications satellites (e.g., weather satellites). The limited domestic investment went into the manned programme, a slimmed-down military programme and navigation systems (GLONASS) – matched by the American Global Positioning System (GPS).

That the programme survived at all was due to the entrepreneurialism of its leaders. They turned the programme around from being the most state-supported in the world to being the most commercial in only a matter of years. By the late 1990s, most of the programme's income came from the launch of Western communications satellites on a fee-paying basis. The manned programme was kept going by American money for joint flights on the Mir space station, selling missions to the European Space Agency and even millionaire tourists. Each design bureau had to fend for itself, but – by 2000 – 87 had commercial or partnership agreements with the West. Against all the odds, the programme did survive. When the American space shuttle burned up in 2003, it was the Russians who kept the International Space Station going. In only four years of the Russian period (1996–9) did Russia slip from its lead position of launching more rockets each year than any other country. By 2000, Russia was once again the world's leading space-faring nation. The 1992 agreement between the Commonwealth of Independent States and the formation of a Russian Space Agency may have been more important than realized at the time, providing a degree of institutional stability.

Like most of its sister and brother bureaux, the Lavochkin design bureau was forced to contract. The company kept working on the military programme, principally the early warning satellites Oko and Prognoz and the optical reconnaissance satellite, Araks. A subsidiary, the Babakin Scientific Research Centre (NITS), was formed to generate foreign business and to organize collaborative enterprises.

The general designer of NPO Lavochkin, Stanislav Kulikov, who had succeeded Vyacheslav Kovtunenko in 1993, was dismissed by the head of the Russian space agency, Yuri Koptev, in August 2003. Kulikov had the misfortune to take up his post just before the Mars 96 mission, but subsequently a number of Lavochkin-built satellites suffered failures: the Cosmos 2344 Araks optical reconnaissance satellite, the Kupon banking communications satellite, two Oko early warning satellites and then a reentry demonstration test. According to news reports in Moscow, ex-Lavochkin Koptev had kept an especially close eye on his former workplace and was distressed by its decline. Lavochkin staff, for their part, criticized Koptev for failing to deliver agreed budgets to the bureau, with the result that future missions had to be delayed or cancelled. Kulikov's position was taken by Konstantin Pichkhadze. But the Russian space industry was a harsh and brutal world, for in March 2004 Koptev was in turn abruptly retired and replaced by General Anatoli Perminov.

Chief designers of OKB-301 (im. S.A. Lavochkin), later NPO Lavochkin

Jul 1937–Jun 1960	Semyon Lavochkin
Jun 1960–Dec 1962	M.M. Pashinin
Dec 1962–Mar 1965	A.I. Eidis
Mar 1965–Aug 1971	Georgi Babakin
Aug 1971–Dec 1977	Sergei Kryukov
Dec 1977–Jul 1995	Vyacheslav Kovtunenko
Jul 1995–Aug 2003	Stanislav Kulikov
Aug 2003	Konstantin Pichkhadze

PLANETARY SPACECRAFT LEAVE FROM BAIKONOUR AGAIN

19th March 2003 marked the sign of better times. That day, a huge, high-winged Antonov 124 cargo plane of the Volga Dnepr company arrived on the runway at Baikonour, with, in its vast hold, the first interplanetary spacecraft to arrive at the launch centre for seven years. On the evening of 2nd June 2003, a Soyuz launcher lifted off into the night sky carrying the ESA's Mars Express and the small British Beagle 2 lander. The European Space Agency paid Russia for the launch.

Through the Soyuz launcher's final stage, the Mars Express launch had a second, stronger connection to the Russian interplanetary programme. The upper stage was called Fregat. Fregat was essentially the CBPS of the UMVL spacecraft and was now adapted by Lavochkin as a versatile, powerful upper stage on Soyuz. With its eight tanks carrying 3,000 kg of fuel, 28 attitude control thrusters, the 1.5 m tall, 3.35 m diameter stage could be used for up to 20 course corrections totalling 877 sec.

Fregat was the unlikely outcome of the blackest day in European rocketry. When Europe's new Ariane 5 blew up soon into its maiden flight in 1996, it took with it an entire flotilla of solar-observing satellites called Cluster, designed to be sent to carefully calculated observing points in high-Earth orbit. Europe's distraught scientists built a fresh version of Cluster, but sought a safer launcher this time. Russia offered the old R-7's latest version, the Soyuz, with an adapted CBPS as an upper stage able to perform the type of complex manoeuvres that were necessary to put the solar observatories into their correct positions. As devised for Cluster, the CBPS offered the possibility of several restarts and could put five-tonne payloads into precise orbits as high as 450 km.

The CBPS was adapted by Lavochkin and the new system acquired the name of Soyuz Fregat. The Soyuz Fregat was launched successfully and put the Cluster series into high orbit in a double launch in July and August 2000. The Soyuz Fregat system was then adapted a second time, now for the Mars Express and its passenger Beagle 2. Europe's first Mars probe entered Mars orbit on 25th December 2003 and over the following two years carried out a highly successful mission. The lander, Beagle 2, was less lucky, being lost and failing to return signals. Two years later, Soyuz Fregat was adapted for a second successful European interplanetary probe, Venus Express.

European spacecraft launched by Russia

Mars Express	2 Jun 2003	Soyuz Fregat
Venus Express	26 Oct 2005	Soyuz Fregat

In a further irony, the American Mars Reconnaissance Orbiter flew to Mars in 2005 on Russian engines. In 1992, Valentin Glushko's old design bureau, now called Energomash, sold its powerful RD-180 rocket to the Americans, who used it for their new fleet of Atlas III and V rockets. The Americans stripped out the engines of their old Atlas, replacing them with just one RD-180, so strong that it had to be throttled back during the ascent to keep vibrations within limits.

REFERENCES

[1] Prismakov, Vladimir; Abramorsky, Yevgeni and Kavelin, Sergei: Vyacheslav Kovtunenko – his life and place in the history of astronautics. American Astronautical Society, *History* series, vol. 26, 1997.

[2] Perminov, V.G.: *The difficult road to Mars – a brief history of Mars exploration in the Soviet Union*. Monographs in Aerospace History, no. 15. NASA, Washington DC, 1999.

[3] Huntress, W.T., Moroz, V.I. and Shevalev, I.L.: Lunar and robotic exploration missions in the 20th century. *Space Science Review*, vol. 107, 2003.

[4] Furniss, Tim: Phobos – the most ambitious mission. *Flight International*, 27th June 1987.

[5] Lemonick, Michael D.: Surging ahead. *Time*, 5th October 1987.

[6] Goldman, Stuart: The legacy of Phobos 2. *Sky and Telescope*, February 1990. For an account of the mission, see Kidger, Neville:
  – Phobos mission ends in failure. *Zenit*, #27, May 1989.
  – Phobos update. *Zenit*, #25, March 1989.
  – Project Phobos – a bold Soviet mission. *Spaceflight*, vol. 30, #7, July 1988.
  For a description of the results, see Results of the Phobos project. Soviet Science and Technology *Almanac* 90. Novosti, 1990; and Zaitsev, Yuri: The successes of Phobos 2. *Spaceflight*, vol. 31, #11, November 1989.

[7] Breus, Tamara: *Venus – the only non-magnetic planet with a magnetic tail*. Institute for Space Research, Moscow, undated.

[8] Surkov, Yuri: *Exploration of terrestrial planets from spacecraft – instrumentation, investigation, interpretation*, 2nd edition. Wiley/Praxis, Chichester, UK, 1997.

[9] Huntress, W.T., Moroz, V.I. and Shevalev, I.L.: Lunar and robotic exploration missions in the 20th century. *Space Science Review*, vol. 107, 2003.

[10] Furniss, Tim: Phobos – the most ambitious mission. *Flight International*, 27th June 1987.

[11] Johnson, Nicholas L.: *The Soviet year in space, 1990*. Teledyne Brown Engineering, Colorado Springs, CO, 1990.

[12] Burnham, Darren and Salmon, Andy:
  – Mars 96 – Russia's return to the red planet. *Spaceflight*, vol. 38, #8, August 1996.
  – On the long and winding road to Mars. *Spaceflight*, vol. 38, #11, November 1996.
  See also: Salmon, Andy: *Mars 96 – the Martian invasion*. Paper presented to the British Interplanetary Society, 1st June 1996; and Mars 96 sera le dernier poids lourd martien. *Air et Cosmos*, 1587, 15 novembre 1996.

[13] Kovtunenko, Vyacheslav M., Kremev, Roald S., Rogovsky, G.N. and Sukhanov, K.G.: *Combined programme of Mars exploration using automatic spacecraft*. Babakin Research Centre, Moscow, 1987.

[14] Kemurdzhian, A.L., Gromov, V.V., Kazhakalo, I.F., Kozlov, G.V., Komissarov, V.I., Korepanov, G.N., Martinov, B.N., Malenkov, V.I., Mityskevich, K.V., Mishkinyuk, V.K. et al.: Soviet developments of planet rovers 1964–1990. CNES et Editions Cepadues: *Missions, technologies and design of planetary mobile vehicles*, 1993, proceedings of conference, Toulouse, September 1992.

[15] Carrier, W. David III: *Soviet rover systems*. Paper presented at Space programmes and technology conference, American Institute of Aeronautics and Astronautics, Huntsville, AL, 24th–26th March 1992. Lunar Geotechnical Institute, Lakeland, FL.

[16] Telegin, Y.: Preparing for the Mars 94 mission. *Spaceflight*, vol. 35, #9, September 1993.

[17] Friedman, Louis: *To Mars via Kamchatka*. Unpublished paper by the Planetary Society.

[18] Kovtunenko, Vyacheslav; Kremev, Roald; Rogovsky, G. and Ivshchenko, Y.: Prospects for using mobile vehicles in missions to Mars and other planets. Babakin Research Centre, Moscow, published by CNES et Editions Cepadues: *Missions, technologies and design of planetary mobile vehicles*, proceedings of conference, Toulouse, September 1992.

[19] Oberg, Jim: The probe that fell to Earth. *New Scientist*, 6th March 1999.

[20] Mitrofanov, Igor: Global distribution of subsurface water measured by Mars Odyssey in Tetsuya Tokano: *Water on Mars and life*. Springer, 2005; Kuzmin, Ruslin: Ground ice in the Martian regolith in Tetsuya Tokano: *Water on Mars and life*. Springer, 2005.

8

Returning to the planets?

Step out onto the soil of asteroids, lift with your hand moon rock, set up moving
stations in space ...

— Konstantin Tsiolkovsky: *Collected works*. Moscow, 1956.

Chapters 3 to 7 showed how the Soviet Union (and then Russia) developed its
unmanned programme to explore Mars and Venus from 1960 to 1996. Chapter 2,
though, left Gleb Maksimov and Konstantin Feoktistov in OKB-1 making plans for a
manned flight to Mars with their TMK-1 and TMK-2 designs, respectively. What
happened to them? What other Soviet projects were advanced for the manned and
unmanned exploration of the planets?

CRITICAL PATHS TO MARS

The main investment in the TMK-1 and TMK-2 project took place over 1959–63.
From then on, it attracted less attention in OKB-1. In a move inconceivable in the
American space programme at the time, the Russians began to fly their designers into
space and the first chosen to go was none other than TMK-2 designer Konstantin
Feoktistov. He made a day-long flight into orbit in October 1964 in the first three-man
spaceship, Voskhod. As for Sergei Korolev, he became ever more involved in the
design of a new manned spaceship, the Soyuz, construction of which began in 1964.
That August, the Soviet government decided to contest the United States in the race to
reach the moon. The N-1 rocket, originally designed for Mars flight, was adapted for
that mission. Korolev was able to give the TMK little more attention in the period till
his early death in January 1966.

The TMK designs continued to attract some low-level interest and attention within OKB-1, renamed after Korolev's death the TsKBEM (and later still, Energiya). The TMK designs were refined again in 1966 and again early in 1969. The May 1966 design was called the KK and was a scaled-down version of the TMK-2, its most notable innovation being the introduction of aerobraking to save energy and fuel [1]. In the 1969 iteration, the crew was down to four; the reactor power had been increased to 15 MW; and the lander had been re-designed – in the face of new information on the thinness of the atmosphere – into a headlight shape.

Apart from the TMK studies, planning for an ultimate manned mission to Mars continued to progress: isolation testing, the biosphere and nuclear engines. These were all what would now be called 'critical paths', the understanding and mastery of which would be necessary for manned flight to Mars.

BIOSPHERE

The idea of biospheres dates back a long time, and closed biological systems were inspired by the ideas of V.I. Vernadsky (1863–1945), after whom one of the most prestigious science institutes in Moscow is now named (the Institute of Geochemistry and Analytical Chemistry). They were self-contained artificial environments on Earth where people were expected to sustain their own air, food and water and dispose of their own wastes, just as they would on a space colony. They were not just a Soviet phenomenon, for the Americans established biospheres in the Arizona desert. When American scientists began to campaign unofficially for a manned mission to Mars (they subversively called themselves 'the Mars underground'), they argued that biospheres were an effective and cheap way of testing the critical paths. In 2000, the Mars Society established a two-floor, 8 m diameter research station or habitat on Devon Island in far northern Canada near an old impact crater, called 'the Hab'.

In 1962, a section of OKB-1 was set up to develop the biosphere, called the NEK or Scientific Experimental Complex and it was put under the charge of Illya Lavrov. Greenhouses were built to grow vegetables and ways were found to regenerate oxygen in closed-cycle systems. In autumn 1964, Lavrov's work was upgraded and he was made head of department #92, charged with continuing to develop plans for a Mars mission. In the event, the biosphere was actually built in the Physics Institute of the Siberian Department of the Academy of Sciences in Krasnoyarsk by Dr Iosif Gitelson.

The first such experiment was called Bios 1. This was quite small, only $12 \, m^3$, but the scientists were able to prove the principle of regeneration of water and air using chlorella, a green algae. The seven-month Bios 3 experiment concluded in Krasnoyarsk in October 1984 with the growing of wheat, tomatoes, cucumbers and other crops in a closed biosystem [2].

Biosphere in Krasnoyarsk

MANOVTSEV, ULYBYSHEV AND BOZHKO FLY TO MARS

Biosphere experiments were designed to test the possibility of cosmonauts growing their own crops and plants to sustain them on long-distance flights. In isolation-testing, the Russians tested the ability of cosmonauts to manage long-duration missions in isolation from the rest of the world. Sometimes the two were combined. Isolation-testing also served the purpose of seeing how cosmonauts could manage on long-duration Earth-orbiting missions and from 1969 Earth-orbiting space stations became an official goal of the Soviet space programme.

For isolation-testing and for research into other aspects of long-duration space-flight, Sergei Korolev assisted in the establishment of the Institute of Medical and Biological Problems (IMBP), a sonorous name that could be calculated not to draw too much attention to itself. There an isolation facility was soon built, partly designed by Korolev himself – formally called the SU-100 but informally *bochka*, or 'the barrel' – and based on the TMK-1 design. It was first tested on 5th November 1967, when three volunteers – team leader Dr Gherman Manovtsev, technician Boris Ulybyshev and biologist Andrei Bozhko – began a year in the *bochka* testing a closed-loop life support system. They grew their own food in greenhouses and recycled air and wastes. A fully self-contained system was not possible and had they depended on the greenhouse alone they would probably have starved. As it was, they lived off canned food.

The marathon of Manovtsev, Ulybyshev and Bozhko was the first and probably the most harmonious of several such isolation tests. The next was a 370-day study in 1986, notable for the stormy arguments between the eight participants. One, Yevgeni Kiryusin, recalled how 'isolation made the simplest irritations turn into a rage.' The third was a 240-day international test in 2000, called SFINCSS (also known as *Sphinx*). This ended in a blazing row concerning the sexual harassment of one

The *bochka*

of the participants, a Canadian woman during a new year party that went wrong. Not only that, but the group divided into two, sealing the hatch off between them for a month until cosmonaut Dr Valeri Poliakov intervened to propose a truce and pacify the warring factions.

The next test, Mars 500, was set for 2007–8. Head of Isolation Mark Belkovsky of the Institute for Medical Biological Problems (IMBP) announced that, with the backing of both the Russian Space Agency and the Russian Academy of Sciences, Russia would recruit 20 volunteers for a simulated 500-day flight to Mars. Twenty volunteers were sought: two groups of six as model 'crews' and a control group, with international participation invited. Funding had been put together from three sources: the federal space budget (160m roubles), the Academy of Sciences and the Ministry of Education and Science. A technical director was appointed, Yevgeni Demin.

He received several immediate enquiries about participating in the experiment, with two cosmonauts indicating their interest. The experiment would mimic flight time outbound, Mars landing and return. The crew would have three tonnes of water and five tonnes of food, generating oxygen by a closed-cycle life support system. The experiment would be carried out in the original *bochka* used for the 1967 experiment, but with the addition of two mock Mars surface modules, bringing the volume of the four-module living space up to 500 m^3. To make the simulation more realistic, there would be longer and longer time delays in the time ground control would take to respond to messages from the *bochka* – 40 min by the time they 'reached' Mars itself. Systems would be put in place to study all the key factors of such a mission: climate, immune systems, toxicology, the nursery, psychology. The experiment would operate a Martian day of 24 hr 40 min. One of the participants would be a doctor and his work would be an important test of telemedecine systems – or as one of the project directors said, 'if they have a fight this time, they'll have to sort it out themselves' [3].

In a controversial move, the IMBP announced that only men would be selected, a point elaborated on by the IMBP director Anatoli Grigoriev, who described a mission to Mars as 'too demanding' for women. Equal rights campaigners abroad at once demanded their countries not participate unless this condition be lifted.

These were serious tests – indeed, early participants received medals for their contributions to science. American scientists have carried out similar experiments north of the Arctic circle in recent years. Between them and –as we shall see – Mir, they form a concrete basis for important aspects of the first Mars expedition.

The *bochka* tests

1967–8	365 days
1986–7	370 days
2000–1	240 days (*Sphinx*)
2007–8	500 days (*Mars 500*)

Anatoli Grigoriev

THE MIR EXPERIENCE

By this time, though, Russia had built up an impressive set of records of real duration missions in orbiting space stations. Starting in 1978 with the Salyut stations, Soviet cosmonauts achieved ever longer records: 96, 139, 175, 185, 237 days in orbit. With the Mir space station, the record was extended twice more. In 1988, two years after the second *bochka*, cosmonauts Vladimir Titov and Musa Manarov flew a 366-day mission on the Mir space station. The ultimate duration flight was carried out by Dr Valeri Poliakov on the Mir station in 1994–5, no fewer than 438 days, the equivalent of the circum-Martian mission envisaged by the TMK-1. By the late 1990s, the Russians had logged the equivalent of several years experience on long-term orbital stations, with the precise effects measured and chronicled.

The Russians made no pretence about the ultimate purpose of the long-duration missions on Salyut and Mir: they were to prepare for a manned flight to Mars. Lengthy missions in Earth orbit of themselves served no useful purpose – indeed, the ideal length of a mission for a resident space station crew turned out to be in the order of six months, the model adopted for the International Space Station. Dr Valeri Poliakov's long-duration mission was explicitly a test for a manned Mars mission and, when he returned to Earth, part of the test was that he could climb out of the descent module unassisted, the rehearsal for a landing on Mars after a long period of travel.

The medical data from his mission were made available to both domestic and American specialists and a video made called *Medicine for Mars*. His experience showed that, with sufficient precautions, it was possible to recover from a very lengthy spaceflight without permanent damage to the body. Dr Poliakov believes that sufficient information was collected on his mission to form the medical basis for a Mars mission. Extreme-duration flights were probably no longer necessary for more tests and if his record were to be broken it would probably be on a real mission to Mars.

The medical precautions to be taken on a manned Mars mission are now well documented from the Soviet and Russian experience of long-duration missions: the importance of two hours exercise every day using different types of machines (e.g., bicycle, treadmill, expanders); the wearing of gravity suits; adequate water supplies; the use of food and vitamin supplements; measures to combat demineralization of the bones, especially the loss of calcium; protection of the immune system; having a doctor on the crew; psychological support from the ground. The effects of zero gravity on the body's function are now well chronicled: atrophying muscles and heart, changed patterns of blood circulation, denser blood, loss of red blood cells, changes in the urinary system, 20–25% calcium loss. But, says Poliakov, the long-term effects of zero gravity are minor and 'Mars gravity, even at 0.37 G, should still restore most of the body's functions' [4].

Countermeasures were an important part of the Russian training for Mars. Russian doctors and scientists came to the conclusion that the disciplined application of countermeasures over long spaceflights would enable humans to function well enough to land on Mars and readapt on their return. If these were adhered to, then humans arriving at Mars would be able to cope with aerobraking, descent, surface activities on Mars and the subsequent ascent. But there were still question marks about the slow, steady rate of bone loss over a 30-month mission and about the accumulated levels of radiation which might not affect mission performance but could provoke cancer later [5].

Cosmonaut Dr Valeri Poliakov

Salyut space garden

Soviet experiments on closed-cycle food generation systems on the Salyut and Mir orbital stations were extensive, starting as far back as 1971 with the hydroponic *Oasis* system on the first space station, Salyut. Plants grown included Chinese cabbage, barley, radishes, mushrooms, onions, cucumbers, tomatoes and even orchids [6]. Mir carried important biosphere experiments. Different types of plants were grown on board (*Oranzheriya*). These included cabbage, broccoli and red mustard. The break-through was with wheat, when on Mir's *Svet* system it was cultivated through the entire cycle from germination to harvesting and subsequent regermination. Wheat was sown on 30th September 1998. Ears appeared on 15th January 1999 and the crop was declared ripe on 1st March 1999, yielding a 500 g harvest. Chickens were even hatched on board. By the end of Mir, the Russians were able to claim success with the regeneration of water, air and liquid human waste. They had been less successful with food and efforts to find ways of reusing solid human waste had made no progress at all [7]. These experiments were continued in the Russian segment of the International Space Station from 1998, with three greenhouse experiments: *Lada*, *Rastennie* and *Svet*.

CRITICAL NUCLEAR, ELECTRIC PATHS

The Soviet Union also made progress with the third critical path, nuclear technologies and electric engines. The most promising was the RD-0410 engine developed by the design bureau K.B. Khim Automatiki (KBKhA) of Voronezh. This began life as far back as 1958 as one of Valentin Glushko's designs, called the RD-410. The Luch design bureau built a test stand for Glushko in the main nuclear testing ground of Obninsk (near Semipalatinsk) under the supervision of the Kurchatov design bureau, the dominant institute of the nuclear industry. Despite the assistance of Luch, Glushko abandoned the project in 1963, but the blueprints were transferred to the

Voronezh OKB-154 Design Bureau for Chemical Automatics or Khim Automatiki. Here it was renamed as RD-0410.

Construction of the engine began in 1965 and tests were eventually carried out from 1970 to 1991. The RD-0410 was a two-tonne hydrogen thermal neutron reactor 3.5 m tall, 1.6 m diameter, with a specific impulse of 910 sec. The engine could fire for up to an hour and be relit ten times. The thrust achieved was 3.5 tonnes, still the highest achieved by nuclear rocket engines [8]. Thirty successful firings were made from 1970 to 1988. Proposals were made for an enlarged version of the RD-0410 called the RD-0411, with 70 tonnes thrust.

Despite having given nuclear engines over to KBKhA Khim Automatiki in 1963, Glushko later returned to nuclear engine design. His OKB-456 now designed a nuclear engine with 17 tonnes of thrust and a specific impulse of 2,000 sec able to generate 200 kW for a Mars mission. In 1983, the bureau revived the studies, publishing the results in 1989. Over ten years, the government approved no fewer than three decrees authorizing construction of nuclear engines, the favoured design being the 11B97 (June 1971: *On work of nuclear rocket engines*; June 1976: *On course of nuclear rocket engines*; and February 1981 *On Herkules inter-orbital tug*).

KBKhA was the only bureau to successfully build and test a Soviet nuclear engine. A working version was to have been tested in Earth orbit and it was slated for launch on a Proton rocket in the second half of 1986. The mission was cancelled after the Chernobyl disaster that April [9].

Smaller engines were also developed and, as already noted, plasma engines were tested on the Mars probe Zond 2. From 1962 Mikhail Melnikov, on instructions from Sergei Korolev, began work on a plasma neutron electric motor of 8.3 kg thrust called the YaERD-2200 and following test bench experiments this became the YaERD-550 motor of 1969 (11B97) with 2.5 kg of thrust [10]. These were developed into a single-block and triple-block configuration of thermionic reactors called the YaE-1 to YaE-3, capable of providing electric thrust of between 6.2 and 9.5 kg, with high-performance specific impulses of between 5,000 sec and 8,000 sec.

Small electric engines were successfully flown at the end of the Soviet period. The first significant test was Cosmos 1066, which flew the Astrophysika SPT-50 electric engine built by the main electric engine builder, Fakel. Later, the famous Arsenal design bureau in Leningrad built the more powerful Plasma A. This was used to provide electrical station-keeping for the satellites Cosmos 1818 and 1867 for over a year.

AELITA – NOT THE FILM, BUT THE REAL THING

Only when the Soviet Union faced defeat in the moon race did a serious reconsideration begin of the original goal of the manned space programme – a flight to Mars. The exact dates and sequence that led to this new goal are unclear, indeed confusing, but the broad outline is now known. Following the successful flight of Apollo 8 around the moon, on 25th January 1969, chief designer Vasili Mishin had a discussion with guidance expert Nikolai Pilyugin about the possibility of a manned Mars mission to take the place of the Soviet lunar effort. Using more powerful versions of the N-1 rocket, they envisaged a sample return mission paving the way for a manned orbit of Mars and then a landing by the end of the 1970s. Already, even though the N-1 had yet to fly, a version with a powerful hydrogen-fuelled upper stage called the N1FV3 was under consideration.

The Council of Chief Designers met in a two-day session on the 26th–27th January 1969 and pursued these options further. Also present was Academy of Science President Mstislav Keldysh and military representatives. The meeting considered whether or not to continue the manned lunar programme; the construction of a large space station in Earth orbit, based on Korolev's TOS designs but now called the MKBS; and a manned flight to Mars. Keldysh favoured an abandonment of the lunar programme in favour of going to Mars, in effect restoring this as the programme's new long-term goal. The outcome was a diffuse one: it was agreed to build the MKBS and then proceed with the Mars plan [11]. They could not bring themselves to make a clean break and cancel the moon programme. The January meeting did agree to continue work on the critical path technologies for Mars – nuclear electric engines and biosphere systems.

The formal decision to make Mars once more a prime Soviet objective in space was issued five months later, on 30th June 1969, as a ministerial resolution by the Minister for General Machine Building Sergei Afanasayev. Echoing Tolstoy's famous film, the resolution referred to it as the *Aelita* programme. Interestingly, it was only a ministerial resolution (#232), not the more imperative type of resolution issued by government and party: though, as the moon contest illustrated, even that did not guarantee a successful outcome.

AELITA: THE NEW TMK–MEK

OKB-1 (now TsKBEM[1]) was invited to update its TMK designs, which it quickly did. Other design bureaux were invited to make proposals. OKB-1's updated TMK project was now renamed – for *Aelita* – the MEK or Mars Expeditionary Complex, the key elements of which were as follows:

- MEK was a 150-tonne complex, requiring two N-1 rockets. One N-1 would carry up the Martian Orbital Complex, the other the Martian Landing complex.
- MEK would make a 630-day mission to Mars, of which 30 days would be orbiting Mars and five days on the surface.
- Electric rocket engines for the main part of the mission, with liquid propellant engines for operations near Mars.

[1] After Korolev's death, OKB-1 was renamed TsKBEM and rival Vladimir Chelomei's bureau TsKBM, the arrangement lasting until the upheaval of 1974. These forms are not used here, since their similarities are so great as to be confusing. The older and more familiar designators are used for convenience.

The MEK looked like a long needle (it was 175 m in length), with the reactors at one end and the crew at the other. The nuclear electric engines would be used to spiral the spaceship out of low-Earth orbit. The crew would have a degree of radiation sheltering. Braking into Mars orbit would take 61 days, with a further 31 days to lower the orbit in preparation for landing. For the landing, three of the crew would go down while three would stay in orbit.

For the return, the nuclear electric engines would burn for 17 days to escape Mars and a further 66 days for transit across the solar system. Three days of braking would be required for capture back into Earth orbit. The final stage of the return to Earth would be in a spaceship similar to the Soyuz design. Models were built.

To send the MEK on its way, chief designer Vasili Mishin had already approved – on 28th May 1969 – the uprating of the N-1 design with liquid-hydrogen upper stages, the new version being called the N1M.

The MEK project received sufficient encouragement for the Institute for Bio Medical Problems in Moscow to construct an experimental complex for the testing of the life support systems involved. IMBP built three modules of 150, 100 and 50 m^3, respectively, including areas for exercise and food preparation.

Vladimir Chelomei with Mstislav Keldysh

AELITA: CHELOMEI'S UR-700M

Apart from OKB-1, the other design bureaux to respond were Vladimir Chelomei's OKB-52 (now called TsKBM) and Michael Yangel's design bureau in Dnepropetrovsk in the Ukraine, OKB Yuzhnoye. Yangel proceeded no further, leaving the field clear for the two traditional rivals, Mishin and Chelomei. Chelomei dusted off the designs of his UR-700 man-on-the-moon mission for an even bigger UR-700M, quite the biggest rocket ever devised by mere Earthlings, weighing up to 16,000 tonnes at liftoff. Chelomei's design went through five variants, three with conventional fuels and two with nuclear – variant 5 having a payload of 1,700 tonnes!

He settled in the end on two possibilities. First was a rocket like his old UR-700, using no fewer than 36 Proton engines on liftoff and diminishing numbers on the upper stages, able to deliver 240 tonnes to Earth orbit. The fourth stage was the most interesting, for he proposed to use the RD-0410 developed by KBKhA Khim Automatiki, then beginning tests in Semipalatinsk. To distinguish his proposal from Mishin's MEK, he called his payload the MK-700 (the name of another variant, the UR-900, was also circulated).

Chelomei's second UR-700M offering was a monster, capable of delivering 750 tonnes to Earth orbit, six times more than America's Saturn V. Two of them would send a 1,500-tonne spaceship to Mars. Although outline models were presented, little further detail is available. He presented his plans in 1970 and offered to run a series of unmanned precursor missions, including recovery of a Mars sample, like project 5NM. As historian Assif Siddiqi correctly points out, Chelomei's plans bordered on fantasy and made absolutely no concessions, in the light of defeat in the moon race, to the declining appetite of the Soviet political leadership for *grands projets* [12]. Chelomei did not focus merely on rocket boosters, for considerable attention was also given to the guidance and navigation systems.

Little of the *Aelita* project leaked out to the West. Occasionally there was a hint. When Soyuz 9 orbited the Earth in June 1970, cosmonaut Alexei Leonov gave an interview in which he outlined how Soviet cosmonauts could fly to Mars on a 17-month journey in 1980 to descend through its thin atmosphere and search for primitive life. Anatoli Blagonravov, chairman of the Space Research Commission of the Academy of Sciences, expressed the hope that the long-duration Soyuz flight would pave the way for a manned flight to Mars [13].

END OF *AELITA*

In the event, *Aelita* had a short life, even as a paper project. Vasili Mishin was overwhelmed by the challenge of getting the N-1 airborne and the development of the space station project, which eventually became known as Salyut. He pulled out of *Aelita*, leaving Mars to Chelomei. Mishin was dismissed as chief designer in May 1974 and retired, carrying on with occasional teaching. In the early 1990s he told the hidden story of the Soviet moon programme and was much missed by historians when he died.

Valentin Glushko – ashes to Venus?

The Military Industrial Commission did include project *Aelita* in the 1971–5 five-year plan, but progress quickly ground to a halt in 1972. An expert commission reviewed the project and found several major problems: it was too expensive (40bn roubles), too big and key issues of propulsion and long-duration missions had still to be solved. By 1973 *Aelita* was no longer an active project, but the paper trail cannot confirm its final end point. Only the 5NM sample recovery project outlived *Aelita* and even then not for long.

Chelomei had a sad end. Ten years later, in December 1984, when getting his Mercedes out of the garage in his dacha (country cottage), the car slipped its brakes and crashed into him on the gates of his dacha. He was taken to hospital but died on 8th December. Although his UR-700 plans were like something out of *Star Trek*, it is not beyond the realm of possibility that some of his earlier designs like the *Kosmoplan* will eventually inspire designs for human missions to the planets. Some of his projects were a happier combination of innovation and successful engineering. The FGB module of the International Space Station comes from an original design by Vladimir Chelomei, so his legacy is very much alive and still circling the Earth.

As for the other character in this drama, Valentin Glushko, he became the chief designer after Vasili Mishin. Combining his old Gas Dynamics Laboratory OKB-486 with Korolev's old OKB-1 into the giant Energiya design bureau, the Energiya/Buran rocket and space shuttle became the *grand projet*, taking the place of Mishin's lunar bases and Chelomei's Mars fantasies. Paralyzed by illness in his final years, he lived to 1989. He was the last of the great designers and his dying wish was that his ashes be brought, one day, to Venus. Maybe they will.

FOLLOWING *AELITA*: ENERGIYA'S NEW DESIGNS

Aelita was the end of manned Soviet Mars projects for 15 years. Over 1986–9, the old OKB-1 – briefly TsKBEM, now Glushko's Energiya – revised its Mars plans, based on the experience of long-duration missions gained on Salyut orbital stations and on the availability of the new Energiya launcher, one much more powerful than the N-1 or the N1M. Energiya may well have been prompted by cosmonaut Konstantin Feoktistov, the original TMK designer of the 1950s. In 1984, he authored *Seven steps into the heavens*, re-making the case for manned flight to Mars. The new Energiya plans envisaged:

- sending a scaled-down version of the Mir space station core module to Mars, with 1.3 tonnes of scientific payload;
- landing on the planet a ready-to-use crew surface vehicle and several rovers;
- assembly of a manned spacecraft, using thin-film solar arrays rather than nuclear power.

The lander design was revised to a cylinder-and-cone lifting body descending horizontally on four legs, with a covered return vehicle in its midst. Whereas the 1986 power plant comprised two independent nuclear reactors, this was later (1989) changed to 200 m wide solar arrays. It is possible that the events in Chernobyl (1986) had something to do with this. Detailed attention was given to the greenhouse systems, now benefitting from several years' practical experience on the Salyut and Mir orbital stations.

Energiya's continued interest in Mars was sufficient to attract the interest of American surveillance, where the director of intelligence ordered an assessment of Soviet manned Martian intentions [14]. The American conclusions were that the USSR intended to mount a manned expedition to Mars after 2000. The intelligence assessment was that the Russians had the infrastructure, experience and motive to do so and were resolving the remaining critical technologies to make such a mission possible. The Americans were clearly impressed with the heavy-lifting capacity of the Energiya rocket, the unmanned programme of Mars exploration outlined for the 1990s and the accumulated experience of long-duration missions on board Mir. They still had more work to do, the Americans observed, in the areas of aerobraking into Mars orbit, nuclear engines and closed ecological systems, but none appeared to be a show-stopper. The Americans estimated the cost of such an expedition at $50bn. The briefing warned the government that the Russians would probably seek American cooperation for such a mission, but that – if this were not forthcoming – the Russians would go ahead anyway, possibly with European partners.

Chernobyl or not, the Russians were not entirely prepared to forsake the idea of nuclear power. In 1994, a new nuclear engine was designed in a joint project between RKK Energiya and the Ministry for the Atomic Industry. This design was called ERTA (*Elektro Raketny Transportniy Apparat*) and was for a 7.5 tonne reactor able to generate 150 kW of energy for up to ten years. In the same year, the Kurchatov Institute designed a Mars expedition around the well-tested RD-0410 engine, clustering together four such engines for the manned Martian expedition.

Different manned approaches to going to Mars were discussed at a meeting of American and Russian officials together in Moscow in September 1998. Participants comprised the US Planetary Society, Energiya and the Institute for Space Research

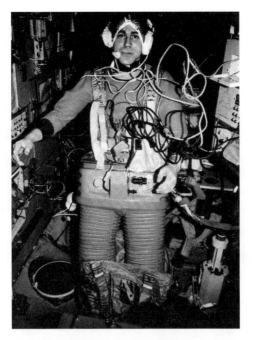

Medical tests on Salyut

(IKI), but it probably took on a surreal quality, for it took place in the midst of the great rouble crash of that autumn.

In 1999, Energiya re-iterated its Mars concept designs, redesigning the lander and switching from one lander to two – one manned, the other logistics. The most significant change related to the method of propulsion. The use of electrical propulsion was reinforced, being adjudged to be the system that was the most reliable, the lowest cost and the easiest to assemble in Earth orbit and would also enable the interplanetary vehicle to be used again. The design bureau observed that, environmentally, this would have the highest level of public support. Accordingly, a 300 N interplanetary thrust engine was envisaged, with a solar plant capacity of 15 MW, a vehicle mass of 600 tonnes, mission time of two years and a crew of six.

An international science and technology committee was set up in 2001 comprising representatives from Russia (eight), the United States (eight) and the European Union (five) to facilitate coordination between national space programmes, in general, and Mars exploration, in particular. On the Russian side, the project involved the federal scientific centres, RKK Energiya, the Institute for Space Research (IKI) and the Institute for Medical and Biological Problems (IMBP).

RKK Energiya quickly contributed to the committee the design of a manned Martian orbital station called MARSPOST (MARS Piloted Orbital Station), which would serve as a base for cosmonauts to drop probes from Mars orbit. MARSPOST was the idea of Leonid Gorshkov of RKK Energiya, who the previous year had combined some of the ideas of the old TMK-1 design for a Mars flyby with the

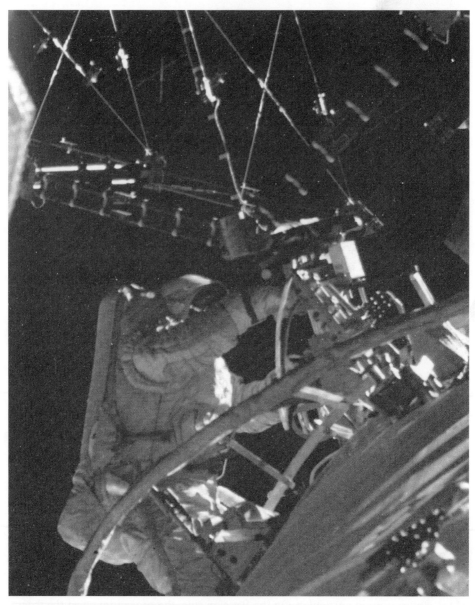

Mir truss – assembly work for Mars

accumulated experience of orbital stations since then. Meantime, the Keldysh Research Centre had already tested a 1:200-scale Martian descent module, designed to
operate with MARSPOST. The descent module would bring cosmonauts down to the
planet, carry a rover for its exploration and enable them to return to the MARSPOST
afterwards. Energiya sketched out a 730-day mission for a crew of six. MARSPOST

would have a mission commander, flight engineer and doctor. The surface expedition would have a pilot, biologist and geologist to spend 30 days there.

Energiya refreshed its Mars designs in 2005. The key elements of its 2005 plan were:

- An interplanetary crew module, shaped like the Mir base block, with a crew of six in six pressurized sections of $410\,\mathrm{m}^3$ volume.
- Large solar electric propulsion array of 30 kW.
- Electric engines, used for acceleration out of Earth's gravity (3 months), outbound (8 months), deceleration into Mars orbit (one month), acceleration from Mars orbit (one month), Earthbound (7 months) and deceleration into Earth's gravitation field (one month).
- Mars lander including, as its top stage, an ascent module. The lander would be 62 tonnes in mass, 40 tonnes on the Mars surface, with an ascent module mass of 22 tonnes, with a capsule of mass 4.3 tonnes.

Soviet/Russian manned flights to Mars: summary of proposals

14 Sep 1956	Start of N-1 design
23 Jun 1960	*On the creation of powerful carrier rockets, satellites, space ships and the mastery of cosmic space 1960–7*
1960–1	TMK designs (TMK-1, TMK-2)
1960	Vladimir Chelomei's Kosmoplan
1969	*Aelita*
1970	Vladimir Chelomei's UR-700M
1986–9	Energiya
1997	Energiya
2001	MARSPOST
2005	Energiya

As was the case with the TMK, work was undertaken on some of the critical path technologies, albeit intermittently. Here, new types of structure were high on the list for testing and, over the years, various structures were tested on the Salyut space stations and the Mir orbital station, with such names as *Ferma*, *Strombus* and *Rapana*. The most sophisticated tubular truss system, the 14 m tall titanium nickel thermo-mechanical *Sofora*, was erected on the Mir space station. A film-type solar array with amorphous silicon 20 µm thick was erected on Mir, anticipating the types of materials required for a Mars mission.

Energiya designed a series of precursor missions to test electric engines, called Modul (Module). Modul was a small experimental spacecraft to be delivered to the International Space Station by the Progress freighter. Once deployed, it would use its electrical engine to raise its orbit to 1,200 km. Over 1998–2001, Energiya built Modul's structure, mechanical units, payload and assemblies, but then funding stopped. In successor missions, Modul M2 would go to a point 1.5m km away from Earth where it would give early warnings of magnetic storms. Modul Mars, the third mission, would go into Mars orbit for two years and return to Earth on electric propulsion. Modul

Mars would study Mars's climate surface and internal structure; carry out remote sensing; and make a global photographic survey.

Modul series

	Module M	Module M2	Module Mars
Mass (kg)	225	960	2,600
Engine	D38	D55	D100
Thrust	0.035	0.05	0.3
Specific impulse	2,080	2,250	3,790
Velocity (km/sec)	0.4	4.5	21

So much for manned flights to Mars and the critical path technology tests. What unmanned programmes were planned during this period?

DZhVS: LONG-DURATION VENUS LANDER

The VEGAs were the last Soviet missions to Venus. Between them, the Venera 9–13 and VEGA landers had characterized the surface, Venera 15 and 16 the planet's main features and the VEGA balloons the atmosphere. As we know from earlier, three more Veneras were constructed, but were reallocated to other programmes. Some missions inspired by VEGA were considered (see *Vesta mission*, below), but did not progress. The Americans concluded the mapping of Venus with the *Magellan* project in 1989. Not for many years did spacecraft return to Venus, although several flew past to use its gravity to assist them on the way to other, more distant, destinations. The next Venus probe was indeed launched from Russia – Venus Express in October 2005 – but it was a European space probe.

The principal contender for a Venus mission during this period was a long-term Venus lander that could operate for up to a month on the surface. This was called the DZhVS in Russian, or long-duration Venus station. The principal aim of the DZhVS was to obtain seismic data about Venus, an experimental objective of Venera 13 and 14 during their short surface transmissions. A second contender was a Venerokhod or Venus rover: conceptual studies were undertaken by VNII Transmash in Leningrad [15]. Pictures show a flat, boxy chassis on wheels crossing a hot, rocky Venusian landscape, with two large dish aerials, one on either side, beaming pictures back to Earth.

The case for the DZhVS was based on the assumption that the accurate measurement of seismic waves could be expected to reveal much about Venus's internal structure, just as similar studies did on Earth's moon. In the original plans by the Lavochkin design bureau, consideration was given to creating artificial seismic events so as to study their subsequent impact. This was a technique developed by the Americans for the Apollo programme, whereby upper rocket stages of the Saturn V were deliberately crashed into the moon to set off seismometers. Apollo 14 astro-

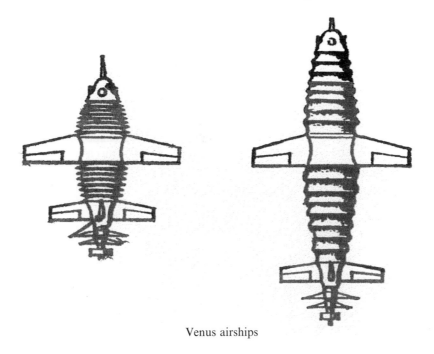

Venus airships

naut Ed Mitchell carried out what was called an 'active seismic experiment' in which a thumper set off 13 shotgun cartridges, the sound waves of which were picked up by a set of geophones in a 91 m straight line.

Firing grenades on another planet was a challenging undertaking, so Lavochkin engineers instead considered having the lander fire an explosive charge separately during the descent, but ensuring of course that the lander reached the surface first so as to be there in time to measure its arrival. This approach had the advantage that it would not require a long-stay lander, for the explosion could be expected not long after the touchdown. This sounded like a good idea, but in the event the problems of separating the delayed action explosive charge proved overwhelming [16].

As a result, scientists decided that the best course of action was to keep a probe long enough on the surface to record a *natural* seismic event. The engineers originally reckoned on a lander that could last five days, but eventually thought that a month might be possible, in the expectation that there would be such a seismic event within that period. The DZhVS had two sections: a small upper section with the traditional Venera equipment (e.g., cameras) and a lower section, using a nuclear isotope generator, with seismic measuring devices and other equipment to measure the changing meteorology of the planet. The project was in the pipeline for a number of years and had the backing of the Institute of Space Research (IKI) in Moscow and its director, Roald Sagdeev. A model DZhVS was tested. In the event, Lavochkin was overwhelmed with VEGA and other work, so the engineers were transferred there and the project withered. It is possible, though, that the design will be dusted off for the latter part of the 2006–2015 federal space plan, should Russia return to Venus.

Other designs for Venus probes were sketched. The success of the VEGA balloons suggested aerostats as a way forward and three aerostat designs were published:

- Dirigible.
- Ring-shaped balloon with large gondola underneath.
- Flat balloon platform in the shape of an oil rig able to drop scientific probes [17].

Science and Life magazine even published designs of a wasp-shaped airship powered by a ducted fan at the rear designed to fly in the clouds of Venus at various altitudes depending on the lifting agent used.

FINAL SOVIET PERIOD PLANS

In the last days of Soviet government, a host of ambitious interplanetary space projects was proposed (e.g., *The USSR in outer space: the year 2005*). In the light of the impending collapse of the system of government, they have a surreal quality now, but they nonetheless represented valid technical approaches to the exploration of the solar system and a historical footnote of what might have been.

In August 1989, the Collegium of the Ministry of Machine Building met in Moscow to plan the Soviet space programme till 2005. The outcome, *Programme for the development of space technology to the year 2005*, was announced at the end of the year. Apart from a solar probe, interplanetary missions were not included at this stage, for they awaited final discussions with the Institute of Space Research. The institute proposed a series of missions to Mercury, the inner and outer planets, and the collegium probably wanted some more time to digest their ambition [18]. Some of the missions had considerable scientific virtue, engineering merit and chances of success: the most outstanding of these was Vesta.

VESTA MISSION

The asteroid Vesta was first declared as a Soviet space objective in Graz, Austria in 1984 and more details were given in March 1985 at the Interplanetary Geology Conference in Houston, Texas. American scientists were stunned by the ambition, scope and detail of the Soviet plans. Vesta is a main-belt asteroid, the second largest after Ceres, orbiting 2.3 AU out from the Sun. It is believed to be covered in basalt and to be scarred by impacts. A landing there could reveal much about the processes of solar system formation and its subsequent history. At this stage, the mission had not even been approved by the government, though this was hoped for that October.

Under the first mission sketch, Vesta was part of a 1992 double mission to Venus, a plan clearly inspired by the success of VEGA and would give France an important role. As the spaceship approached Venus, it would split, the French part setting off for asteroid Vesta, in much the same way as the interception of Halley was organized. The Soviet probe would drop a kite (not a balloon) that would fly in the Venusian atmosphere for a month, lowering an instrument package on a 20 km long cable below to sample the atmosphere at different levels. Meantime, the lander would turn its cameras on at 18 km and film the rest of the descent down to the surface. Once there, surface experiments would be carried out by an X-ray fluorescent spectrometer, X-ray spectrometer and gas chromatography mass spectrometer.

The French part of the mission would pass asteroids Kalypso and Tea before approaching Vesta, where a small lander would be dropped. Seven instrument packages were outlined: camera, infrared spectrometer and radiometer on the main spacecraft and on the lander a camera, gamma-ray spectrometer and X-ray spectrometer for surface mineral and chemistry experiments. When they heard all that, no wonder the hardened Americans gasped [19]. The mission underwent further refinement in the course of the year [20].

To prepare for the mission, in September 1986, Soviet astronomers using the 60 cm telescope high in the Tien Shan Mountains 1,500 m above sea level in Kyrgyzstan began a survey of the asteroid belt so as to better target the mission. They made multiple observations of Vesta, which they determined to be 580 km across, nearly spherical and revolving on its axis every 5.2 hours.

In 1986, the Vesta mission was reconsidered. A new trajectory was proposed, with a gravity assist at Mars, with consideration also being given to a sample return. The 'Martians' now appeared to be winning over the 'Venusians', for Venus was now dropped from the flight profile.

In December 1986, the Vesta mission was finally redefined via Mars. Discussions were under way for a French participation in the order of 15–20%. This reached quite an advanced stage of planning: launch dates were even set. The mission would unroll as follows:

- Launch by Proton rocket of two spacecraft from Baikonour to Mars (23rd September and 12th December 1994).
- Arrival at Mars on 31st May 1995 and 15th January 1996, respectively.
- Russian spacecraft enter Mars polar orbit of 200×800 km.
- Russian–French asteroid spacecraft use Mars for two gravity assists (mid-1995 and mid-1997) for tour of small bodies of the solar system.
- Targeting of eight asteroids: two large – Vesta (575 km), Hestia (165 km) – and up to six small (James, Felicitas, Mandeville), including possibly Comets Tritton or Lovas over 1995–9.

These would be large spacecraft of the UMVL type. The main Russian spacecraft, weighing three tonnes at Mars arrival would enter Mars orbit where it would bring a 200 kg payload. This would comprise camera, radar altimeter, micrometeoid detector and spectrometers (gamma, mass, infrared, ultraviolet). Three descent modules were under consideration, each in the order of 100 kg weight: lander, penetrator or balloon.

For the Russian–French spacecraft, Mars would be used for gravity assist. The upper half of the spacecraft would see the main French contribution. Here, the French proposed to adapt a 1,500 kg Eurostar-type communications satellite of the version used for Inmarsat, with 3 kW solar panels. This would carry 750 kg of fuel so as to manoeuvre to the various targets, with cameras, dust detectors, radio altimeter and infrared spectrometer. As the French probe approached each of the large asteroids, at a closing velocity of 3.3–3.6 km/sec, it would fire a 500 kg penetrator between 500 km and 1,000 km out toward the two largest asteroids. Each penetrometer would carry a spectrometer, accelerometer, magnetometer and cameras and would transmit data back to Earth through the French probe. The French hoped that the European Space Agency would contribute €100m to the French Space Agency's estimated €166m cost of the French part of the project. During the interception of Vesta, measurements would be made of its size, shape, volume, density, period of rotation, morphology and mineral composition.

A year later, in November 1987, the Vesta mission was put back to 1994, but was

French balloons over Mars in early 1990s' plan

still portrayed as the flagship interplanetary mission of the 1990s. The French appear to have been slow to make a decision to join, but the Russians hinted that they would go ahead anyway, possibly using the asteroid spacecraft to retrieve samples [21]. Whatever their supposed hesitation at this stage, the French Space Agency, CNES, together with the European Space Agency (ESA) completed a full study of the mission the following year [22] for a tripartite mission managed by CNES, ESA and Inter-cosmos. This redefined the mission yet again and pushed the launch back even further, to February 1997. Now there would be two identical spacecraft, each having a French converted communications satellite on top, the Russians in effect providing the mother spacecraft.

Already, financial restrictions had forced the Russians to reduce the number of launchers available from two Protons to one, with the identical spacecraft sent up on just one Proton. Each would use Mars for gravity assist before beginning a seven-year tour of the solar system's asteroids, but the the Vesta probe would make four close flybys of Mars, coming as near as 300 km, where high-resolution imaging would be carried out and a radar would make a temperature profile of the atmosphere. The idea was that one of the two missions would focus on asteroid Vesta. A series of trajectories and options were worked out for as many as seven secondary asteroids (Horembh, Iris, Broederstrom, Ron Helin, Roxane, James, Hestia) and two comets (Bus and Dutoit–Neujmin–Delporte), the comets lying between 1.8 and 1.9 AU out from the Sun). It was planned to slam penetrators into two bodies: the stony asteroid Iris and the carbonaceous asteroid Hestia. A substantial amont of cruise science was envisaged. The penetrators would carry nine instruments, including a camera on top which would peer out from the fin to take a close-range picture of the surface, a seismometer, thermal probe, spectrometers and soil sampler. Tracking would be a European responsibility, to be undertaken by a worldwide network including Weilheim in Germany, Usuda in Japan and Algonquin in Canada.

Despite having the benefit of a mission study, it is not known how much progress the mission made. The scientists taking part were probably very much involved in the Phobos mission over 1988–9 and its subsequent unravelling. When a new, post-Phobos launch schedule was published only a year later in 1989, Vesta had disappeared from the manifest altogether. At one stage, there was a proposal, with an air of desperation about it, that the Phobos 3 flight model be adapted for the Vesta mission. Vesta re-appeared briefly and for a final time on the mission schedule issued by the Russian Space Agency in September 1992, by which time it had acquired a new name, Mars Aster. Still with a 1996 start, it would fly past Mars in October 1997 and then begin a tour of the asteroids, starting with Fortuna (September 1998), Harmonia (November 1998), Vesta (June 1999), Halej (October 1999) and concluding with Juewa in July 2001.

Vesta and Mars Aster disappear from the record at this stage but, like *Aelita*, they do not seem to have been formally or officially buried. In the end, they were overwhelmed by the financial crisis that overtook all the later Mars missions, leaving the unfortunate Mars 96 as the sole survivor. In the end, NASA achieved the first landing on an asteroid, Eros, with the NEAR mission in 2001 and later approved a *Discovery*-class mission to fly to asteroids Vesta and Ceres.

PUTTING MARS BACK TOGETHER AGAIN

At least a year before Mars 96 failed, work had stopped on all the other Russian Mars probes: Mars 98, sample return missions, Mars Aster, with all their various permutations of orbiters, rovers and balloons. The French, twice frustrated in attempts to put together balloon projects with the Russians, took the decision that they were unreliable partners. French Space Minister Claude Allegre was quoted publicly as saying that the Russian Mars programme was no longer credible and in 2000 signed a much-publicized programme for joint robotic exploration with the Americans, involving a network of surface stations (Netlander) and a sample return mission. Despite this, the main French contractor made it clear that it still favoured the Lavochkin design bureau to develop the surface stations. Within a few months, the headstrong flirtation with the United States fizzled, little more was heard of the joint programme and a new minister took his place.

Deserted by the French, the Russians made a last approach to their old rivals, the Americans. On 17th June 1992, the Russian Federation and the United States signed *Agreement on the exploration and use of space for peaceful purposes*, with the details to be worked out later by a joint commission of the Russian Prime Minister Viktor Chernomyrdin and the American Vice President Al Gore (the Chernomyrdin–Gore Commission). A year later, the Russians and Americans effectively merged their two space station programmes (Mir 2, Freedom) into the International Space Station, eventually launched in 1998. Could this cooperative spirit extend to Mars missions?

At one stage, the prospects looked bright. Together, the Americans and Russians explored a project called Mars Together. When Russia cancelled Mars 98, Mars Together was envisaged as a replacement project. Under Mars Together, the Russians would supply an 8K78M Molniya launcher and a Mars lander of 320 kg, equipped with a 95 kg Marsokhod carrying 12 kg of experiments able to roam hundreds of kilometres across the planet. The American orbiter would carry instruments duplicating those lost on Mars Observer, principally a gamma-ray spectrometer and an infrared radiometer. Ultimately, this project collapsed too, though the precise circumstances of its demise are not known and NASA pursued its own programme through the series of missions flown over 1997–2003: Mars Global Surveyor, Pathfinder, Odyssey and the two Mars exploration rovers.

RETURN TO PHOBOS? PHOBOS GRUNT

The idea of the Mars sample return has always been a strong theme in Russian Mars exploration and it featured repeatedly in the on–off plans of the 1990s. The influence of Gavril Tikhov proved to be a long one. Years after his death in 1960, Russian scientists continued to explore deep into the Siberian permafrost to determine at what depth micro-organisms could survive. On Earth, they found them down to a depth of 10–15 m, with water of course playing an important life-sustaining role. As a result, Russian plans for the exploration of Mars have always emphasized the importance of drilling down, where Martian micro-organisms would be protected from radiation,

Soviet period sample return mission

possibly in subsurface water, hence TERMOSCAN. As scientist Anders Hansson says: 'Surface samples only tell you about the biological *present* and are not very useful. The real discoveries lie in drilling down to the subsurface water where you find the biological *past*' [23]. So the idea of a sample return flight, first from Phobos, later from Mars's surface itself, could never be kept off the agenda for long.

Earlier studies had shown just what a difficult undertaking it would be to recover samples from Mars. But what about one of its moons? In 1999, the Russian Academy of Sciences, through the Institute of Space Research (IKI) and OKB Lavochkin began a feasibility study of recovering rock samples from the moon Phobos and invested an initial 9m roubles in the project. Called Phobos Grunt (literally in Russian, 'Phobos Soil'), the probe was similar to but smaller than the failed Phobos missions of 1988–9 and would follow a similar mission profile. Indeed, several years earlier IKI had floated the notion of converting the flight model of Phobos 3 for such a mission.

An important underlying assumption of a new mission would be that the Proton rocket was too expensive and that the less powerful 8K78M must be used instead. Like Mars 96, the Phobos Grunt's own engine would have to complete the transfer out of Earth orbit, but this time, in a radical departure in profile, electrical engines would be used. The spacecraft would use nine SPT-100V electrical engines originally developed by the Fakel Company for station-keeping for communications satellites. These electrical engines, with a thrust of 130 mN and an ejection speed of 22 km/sec, would use 425 kg of xenon to speed the probe on a slow, 475-day trajectory toward Mars.

After entering Mars orbit, the 2,370 kg probe would close in on Phobos in the course of a three-week rendezvous. The electric engines would be used for the main manoeuvres in Mars orbit, but ordinary chemical engines would be used for the very final stage of closing in on Phobos, following manoeuvres similar to those carried out or intended in 1989. A couple of days after scooping up 170 g of rock samples from Phobos, a small capsule with the samples would be fired toward Earth, to be recovered 280 days later in Russia.

The Russians continued to promote Phobos Grunt over the years and in 2003 issued a revised specification for the mission. The launcher proposed was the now proven Soyuz Fregat. The electrical propulsion system, called the SPT-140, was upgraded to three thrusters using 4.5 kW power. The spacecraft looked like the Phobos mission of 1988, but now with very large solar wings for electrical power.

Under this proposal, Soyuz Fregat would put the Phobos Grunt in a high-Earth orbit of of 215 × 9,385 km. At this stage, the Fregat stage would be dropped and the solar electric power take over to spiral the spacecraft, now 4,660 kg, out of Earth orbit and toward Mars, the huge solar panels being dropped at Mars orbital insertion. Once in Martian orbit, the spacecraft would follow the path of Phobos 2 in 1989, entering an orbit that passed Phobos every four days. Small rockets would be used to press the spacecraft onto the surface and anchor it in the low gravity. The entire spacecraft was designed to land and remain on Phobos for some time, powered by solar panels. A Luna 24 type drill would be used to collect rock, the ascent rocket being fired from the top for a Luna-style return to Earth. The Russians believed that they could build a spacecraft with a surface payload of 120 kg, which would be substantial and attractive

Old target, new target: Phobos

from the point of view of international participation. From 2004, the Russians began to sound out, informally, the prospects of such international cooperation.

For many years, Phobos Grunt looked like another paper project that was never going anywhere. It was a repeated theme on the wish list of Soviet planetary scientists and engineers. Their fixity of purpose eventually paid off, for when the ten-year Soviet space plan for 2006–2015 was published in summer 2005, Phobos Grunt was one of only two new flagship projects (the other was the new space shuttle, the Kliper). Launch was set for October 2009. A project manager, Igor Goroshkov, was appointed. Detailed sketches of the revised Lavochkin design were published. There were further revisions to the 2003 plan, though the flight profile was much the same. The two-tonne spacecraft was now a low, squat, neat polygon on four legs, with a new type of drilling rig at the side, with a long tube for vacuuming the soil sample into the recovery cabin. The lander would work on Phobos for a year after the return stage was fired back to Earth, equipped with about 20 science experiments provided by Russia and the international science community. The return stage was a small, box-shaped module with a solar panel on one side, the ball-shaped sample recovery cabin fitting snugly in the middle. The return stage would lift off the lander, enter Martian orbit, orientate itself, fire to Earth, transit for eleven months and then release the small sample recovery cabin for its plunge into the atmosphere. The small size of the spacecraft, benefitting from recent advances in electronics, miniaturization and materials, was a dramatic contrast to the size and complexity of Sergei Kryukov's project in the second half of the 1970s [24].

China joined the project the following year. China was already planning its first moon probe – the Chang e – and had expressed an interest in sending small spacecraft to Mars. In 2006, it was learned that Phobos Grunt would, once it arrived in Mars orbit, deploy a small Chinese subsatellite, the Yinghuo 1, into a $800 \times 80,000$-km equatorial orbit. Yinghuo, weighing 120 kg, would be the first subsatellite of Mars and would study Mars's atmosphere and ionosphere with Phobos Grunt. This was a clever concept, getting China to Mars much sooner than would have otherwise been the case, opening up the scope for simultaneous observations by two satellites and furnishing some funding for Russia.

FIRE AND ICE

Phobos Grunt was the only fresh project that emerged from the Russian period with any prospect of fulfilment, but it was not the only project considered during the period. Back in 1987, the Institute for Space Research (IKI) put forward its wish list for planetary exploration projects. These were:

- Project Koronas, a flight out to Jupiter curving back to the Sun with a flyby at between 5 and 7 solar radii, set for 1995.
- Mission to Jupiter and Saturn, 1999, with a lander and atmospheric balloon probe to Titan. The balloon would actually drop the lander on the surface before flying back up to its operational altitude. The lander would carry 50 kg of equipment and be designed to survive a pressure of 1,000 atmospheres. The atmospheric probe would be a balloon with 5 kg of experimentation to circle between 5,000 and 8,000 m high in the atmosphere for about ten days.
- Mercury mission for 2002–3, using a Venus swingby, with the main spacecraft entering Mercury orbit and a lander descending to the surface.
- The Vesta and other Mars missions discussed earlier [25].

As noted earlier, in August 1989 the Collegium of the Ministry of Machine Building approved the *Programme for the development of space technology to the year 2005* including a solar probe (Solnechny Zond), stating that further discussions had still to take place with the Institute for Space Research which would make announcements in due course. Taking the IKI wish list into account, it added a Mercury probe, including a lander, for 2001–3. The Saturn probe does not appear to have won collegium support: indeed, the proposal was astonishingly similar to the American–European *Cassini–Huygens* mission later launched in 1997, which achieved stunning success eight years later.

The solar probe made some progress. A design study was concluded in 1990. The probe, renamed *Tsiolkovsky*, would fly out to Jupiter and then take a curving trajectory back in toward the sun, which it would pass at the very close distance

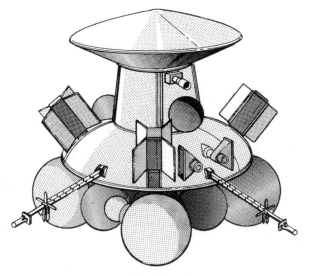

Tsiolkovsky probe

of 4m km. *Tsiolkovsky* would use the UMVL design, the CBPS carrying the largest possible fuel load to accelerate the probe out toward Jupiter. The spacecraft itself would weigh up to 1,200 kg and carry radio-isotope generators for a mission of up to five years.

Tsiolkovsky took a second turn four years later. It became part of the 1992 agreement and the work of the Chernomyrdin–Gore Commission. In April 1994, NASA and the Russian Space Agency decided to draft a concept for two contrasting missions which included ideas from the *Tsiolkovsky* study. The first, called *Plamya* ('fire') was to send two probes, one Russian to within ten radii of the Sun and an American one to four radii; the second, called *Lyod* ('ice') was to send an American probe to reach the only unexplored planet of the solar system, far distant Pluto, with a Russian capsule. Launched by the Proton rocket, the two *Plamya* probes would travel out to Jupiter before swinging back toward the sun on a 800-day journey. The Russian probe was to be 350 kg in mass, with 35 kg of experiments; the American 200 kg, with 22 kg experiments. As they passed by the sun, they would film its disk with a variety of instruments. *Lyod* was a ten-year mission, using the smaller 8K78M Molniya rocket. The 85 kg *Lyod* ice probe was expected to pass Pluto at 15,000 km and its moon Charon at 5,000 km, sending down a 10 kg Russian capsule to fly close by or even impact [26].

In much the same way as Mars Together, *Plamya i Lyod* did not progress as a joint project. It is possible that it was made partly redundant by the success of the European–American *Ulysses*, which passed through the Jovian system in 1992 and then under the sun in 1994. NASA dropped the Pluto mission, though in a reversal of normal roles it was eventually reinstated by the Congress and finally launched in 2006. On the Russian side, something was salvaged too, for an Earth-orbiting solar observatory, Koronas, was launched in 1994.

Russia returned to the idea of a Jupiter mission in 2004. Following the spectacular tour of the Jovian system by the American spacecraft *Galileo*, there was renewed interest in further missions to explore its large icy moons. The administrator of NASA, Dan Goldin, promoted a programme called Prometheus to develop a nuclear power capability to support such a mission. A lead project on the NASA wish list for years was called JIMO, or Jupiter Icy Moon Orbiter.

Russia proposed its own version of JIMO in 2004. The proposal, led by the Keldysh Centre and the Lavochkin design bureau was for an orbiter of the moon Europa. Using the new, heavy-lift, Russian launcher, the Angara 5, the 9.6 tonne spacecraft would leave Earth orbit on an eight-year journey, assisted by an electronuclear engine with a power of 100 kW and specific impulse of 4,500 sec. The spacecraft would have a payload of 1,250 kg, comprising a one-tonne radar using 30 kW of power, designed to study the coverage and composition of the ice of Europa down to a depth of 70–80 km [27]. The status of the proposal remains unknown. It may have been intended for the new Russian space plan, but does not seem to have progressed.

The final deep space project deriving from the end of the Soviet period was an experimental solar sail project, Regata. This concluded design studies, but hardware does not seem to have been built. Regata was announced at a conference in Frunze in early 1990. Regata was to have a 9 m solar sail, behind which would be a 530 kg platform including a 230 kg payload. The intention was to place a fleet of the spacecraft at the Lagrangian points around the Earth where it would carry out observations of the sun and the interplanetary environment. A version, Regata Astro, was designed to fly out to Mars to precisely measure the planet's orbit, while another was under consideration for interception missions to asteroids, comets and small bodies of the solar system. The project was an early victim of Russia's rapidly shrinking science budget and never got further than a paper project.

Projected missions, 1987–2007

Name of mission	Objective
DZhVS	Long-duration Venus lander
Vesta	Asteroids via Venus, later Mars
Mars Aster	Asteroids via Mars
Mars Together	Mars lander, orbit, balloon, Marsokhod
Phobos Grunt	Soil recovery from moon Phobos
Tsiolkovsky	Jupiter and Sun
Korona	Jupiter and Sun
Solnechni Zond	Sun
Plamya i Lyod	Jupiter and Sun/Pluto
Jupiter, Saturn, Titan	Jupiter, Saturn, Titan (with probe and balloon)
Regata Astro	Mars
JIMO	Europa

KEEPING THE DREAM ALIVE

As one may imagine, the 2001 International Science and Technology Committee did not have sufficient resources to organize a mission to Mars. Still, within its existing resources, it was able to provide a grant for its Russian members to put together a 30-volume publication on their state of knowledge of Mars and scenarios for landing there. The subsequent MARSPOST proposals show that – though finances, governments and régimes may come and go – Russian scientists have still not abandoned the idea of landing on the planet Mars. Professor Nikolai Rynin's encyclopaedia of 1927 was first written when the Soviet Union was still devastated by civil war and parts of the country were starving, so hard times were never considered in themselves a sufficient reason for giving up. The planet continues to be a focal point in Russian thinking.

During the 1990s, the Russians planned a series of projects for the further exploration of Mars and the solar system. Most owed their origins to the end of the Soviet period and were conceived when horizons and finance appeared limitless. The most persistent of these early missions was Vesta, which at one stage appeared to be a certainty. Many of the other missions planned for the period would have added considerably to our knowledge. Balloons have yet to be flown on Mars or to the atmosphere of Titan. The Russian Marsokhods would have travelled much farther than the American Sojourner, Spirit or Opportunity. Although the first asteroids have now been visited (Eros, Itokawa) and visits to others are planned (Ceres, Vesta) they fall short of the full tour projected 20 years ago.

Remarkably, the Russians kept alive the TMK studies developed by Sergei Korolev, Mikhail Tikhonravov, Gleb Maksimov and Konstantin Feoktistov in the late 1950s. It is now possible to trace, as Energiya has done, how the mission concepts evolved over the intervening years [28]. Plans now are more modest in scope, respectful of the budgets likely to be available and, probably for environmental reasons, reoriented around the use of solar electric propulsion. Even in their own limited way, the Russians have kept going their development of the critical path technologies: isolation-testing, biosphere, truss and structure technology, electric engines. The 30-volume assembled knowledge of Mars and the techniques for getting there will ensure that the scientific and institutional memory of the period will be available for future generations.

Even as the 50th anniversary of the first Sputnik approached, Russians designers continued to hope for missions to the planets. The Lavochkin design bureau continued to work on fresh designs. Once Phobos Grunt is successful, a Mars Grunt would follow in around 2013–2016, taking samples from the Martian surface and sending them back to Earth like the lunar sample return missions had done in the 1970s [29].

REFERENCES

[1] Mark Wade has made a series of studies of Soviet and Russian manned plans to fly to Mars, including the nuclear technologies associated with them and these are cited here as Wade, Mark:
 - TMK-1;
 - TMK-E;
 - MPK;
 - KK;
 - Mavr;
 - MEK;
 - MK-700;
 - Mars 1986;
 - Mars 1989;
 - ERTA;
 - Mars 1994;
 - MARSPOST *Encyclopedia Astronautica, http://www.astronautix.com*, 2005.

[2] Hansson, Anders: *The Mars environment in Russia*. Paper presented to the British Interplanetary Society, 12th June 1993.

[3] Parfitt, Tom: Spaceflight is hell on Earth. *The Guardian*, 8th September 2005. See also: Oberg, Jim: Are women up to the job of exploring Mars? *MSNBC*, 11th February 2005; Phelan, Dominic: Russian space medicine still aims for Mars. *Spaceflight*, vol. 46, #1, January 2004; Zaitsev, Yuri: Preparing for Mars – a simulated manned mission to the red planet is about to begin. *Spaceflight*, vol. 47, #1, January 2005.

[4] Poliakov, Dr Valeri: Remarks made at presentation in British Interplanetary Society, London, 22nd May 2002; Kozlovskaya, Inessa and Grigoriev, Inessa: *Countermeasures*. Paper presented at International Astronautical Federation conference, Bremen, Germany, 2003.

[5] Orlov, Oleg and White, Ron: *The medical challenge*. Paper presented at International Astronautical Federation conference, Bremen, Germany, 2003.

[6] Harland, David M.: *The story of the space station Mir*. Springer/Praxis, Chichester, UK, 2005.

[7] Gitelson, Josef: *Mars – to go there, we start here*. Paper presented at the International Astronautical Federation conference, Bremen, Germany.

[8] Lardier, Christian: CADB dévoile le moteur nucléaire RD-0410. *Air et Cosmos*, vol. 1571, 21 juin 1996.

[9] Rachuk, Vladimir: Best rocket engines from Voronezh. *Aerospace Journal*, November/December 1996; Hansson, Anders: *Russian nuclear propulsion*. Paper presented to the British Interplanetary Society, 25th May 2002.

[10] Les moteurs nucléaires de l'ère soviétique. *Air et Cosmos*, 1810, 21 septembre 2001.

[11] Zak, Anatoli: *Martian expedition, http://www.russianspaceweb.com*

[12] Siddiqi, Assif: *The challenge to Apollo*. NASA, Washington DC, 2000.

[13] Angus McPherson: Mars by 1980? Russia shakes the west. *Daily Mail*, 3rd June 1970; Russian hope of Mars flights. *The Times*, 26th June 1970.

[14] Central Intelligence Agency (1989): *Soviet options for a manned Mars landing mission – an intelligence assessment*. Director of Intelligence, CIA, Washington DC.

[15] Kemurdzhian, A.L., Gromov, V.V., Kazhakalo, I.F., Kozlov, G.V., Komissarov, V.I., Korepanov, G.N., Martinov, B.N., Malenkov, V.I., Mityskevich, K.V., Mishkinyuk, V.K.

et al.: Soviet developments of planet rovers 1964–1990. CNES and Editions Cepadues: *Missions, technologies and design of planetary mobile vehicles*, 1993, proceedings of conference, Toulouse, September 1992.

[16] Hendrickx, Bart: *Soviet Venus lander revealed*, Friends and Partners in Space posting, 30th August 2001.

[17] Yumansky, S.P.: *Kosmonautika – Segondniya i zavtra.* Prosveshchenie, Moscow, 1986.

[18] IKI (Institute of Space Research): *The Soviet programme of space exploration for the period ending in the year 2000 – plans, projects and international cooperation. Part 2: The planets and small planets of the solar system.* Institute of Space Research, USSR Academy of Sciences, Moscow, 1987.

[19] Beatty, J. Kelly : The planet next door. *Sky and Telescope*, 1985; see also Covault, Craig: Soviets in Houston reveal new lunar, Mars, asteroid flights. *Aviation Week and Space Technology*, 1st April 1985.

[20] Lenorovitz, Jeffrey M.: France designing spacecraft for Soviet interplanetary missions. *Aviation Week and Space Technology*, 7th October 1985; see also Langereux, Pierre: Les quatre sondes Franco-Soviétiques VESTA vont explorer Mars et les petits corps. *Air et Cosmos*, #1117, novembre 1986.

[21] Furniss, Tim: Countdown to cooperation. *Flight International*, 5th December 1987.

[22] European Space Agency (ESA) and Centre National d'Etudes Spatiales: *VESTA – a mission to the small bodies of the solar system: report on the phase A study.* Paris, 1988.

[23] Hansson, Anders: *The Mars environment in Russia.* Paper presented to the British Interplanetary Society, 12th June 1993.

[24] Popov, G.A., Obukhov, V.A., Kulikov, S.D., Goroshkov, I.N. and Upensky, G.R.: *State of the art for the Phobos Soil return mission.* Paper presented to 54th International Astronautical Congress, Bremen, Germany, 29th September–3rd October 2003; Ball, Andrew: *Phobos Grunt – an update.* Paper presented to the British Interplanetary Society, 5th June 2004; Craig Covault: Russian exploration – Phobos sample return readied as Putin's government weighs Moon/Mars goals. *Aviation Week and Space Technology*, 17th July 2006.

[25] Furniss, Tim: Countdown to cooperation. *Flight International*, 5th December 1987; Johnson, Nicholas L.: *The Soviet year in space, 1990.* Teledyne Brown Engineering, Colorado Springs, CO, 1990.

[26] Lardier, Christian: Les nouveaux projets de la NPO Lavochkine. *Air et Cosmos*, 18 avril 1997, #1609.

[27] Lardier, Christian: Le Jimo, version russe. *Air et Cosmos*, #1955, 22 octobre 2004.

[28] Present Russian Mars plans: *http://www.energia.ru.english/Energiya/mars*

[29] Kopik, A.: Big plans at NPO Lavochkin. *Novosti Kosmonautiki*, vol. 15, #10, 2005.

9

The legacy

Humankind's conquest of the solar system is of crucial importance because it would give us, later, the vision for the proper management of our home planet.

– Yuri Kondratyuk, AKA Alexander Shargei, 1928

The inclusion of Phobos Grunt in the Russian space programme for 2006–2015 means that Russia is, after a gap of more than ten years, set to return to the planets. Like the programmes of the Soviet period, the mission has its ambiguities. Phobos Grunt represents a victory for the 'Venusians', insofar as it does not compete directly with the United States or any other country, for no such mission is planned by them. Having said that, the timing puts it just ahead of the schedule of when the Americans do hope to get samples back from Mars. But in getting samples back from a Martian moon, the Russians will be able to claim, for the first time in a generation, a new, satisfying planetary 'first'.

The Russian programme for the exploration of the planets has, as we have seen, had a long history, tracing its roots back to Konstantin Tsiolkovsky in tsarist times and other great theorists spanning the tsarist period and the revolution, like Friedrich Tsander and Yuri Kondratyuk. Mars was a central, focal, reference point in the work of many, especially for Tsander; his influence was durable, for Sergei Korolev continued to invoke his name into the 1960s. These theorists were no mere idle dreamers, for nearly all of them were also involved in practical rocketry and attempted to overcome the most basic problems of propulsion, electrics, firing systems, tanking, valves and guidance. Their work was well assisted by the promotional works of Yakov Perelman and Alexei Tolstoy, who helped to embed the idea of flight to Mars within the popular consciousness, achieving an imprint which, despite the Stalinist period, never entirely disappeared. Their works of fiction were given an additional edge by the attention given to Mars and Venus by Gavril Tikhov, who called himself an astrobotanist, but would now be considered the first of the exobiologists. His theories of the survival of life forms on Mars, similar to those to be found in the permafrost of Siberia and his drawing of colourful blue plants thriving in the watery swamps of Venus, captivated a generation of astronomers and space scientists.

Once they had achieved the launching of Sputnik, Mikhail Tikhonravov and Sergei Korolev turned to the objective of Mars with breathtaking speed. Korolev hoped to send probes to Mars and Venus only months after Sputnik was launched. Nevertheless, even at a time when he was preparing for the daunting task of man's first flight into space, Korolev busied himself with the design of the first series of Mars and Venus probes (1MV), organized the construction of the Yevpatoria tracking station and pressed the Institute of Applied Mathematics into calculating interplanetary trajectories. The first Mars and Venus probes had all the ambition typical of Korolev's space plans. At a time when the Americans were mulling over flyby missions some time in the future, Korolev was already building landers. Although Tikhov died in 1960, Korolev's Mars landers were still expected to find life, while the cabins of the first Venera probes were built to float on the blue lily ponds of Venus.

It was only years later that we learned that Korolev was simultaneously proceeding with constructing the rocket that was to fly the first cosmonauts to Mars. The N-1 rocket, whose blueprints dated to September 1956, was originally built around a Mars mission, not a lunar landing. Had events not taken the turn they did, the N-1 might have proved ideal for the assembly of the first Martian expedition in Earth orbit, rather than, in its larger form, a less-than-successful moon rocket. Korolev set his best and brightest designers – Tikhonravov, Maksimov, Feoktistov – to design the first Mars mission, the *Tizhuly Mezhplanetny Korabl* (TMK). Reminding them of Tsander's injunction ('We *shall* fly to Mars'), he pressed them like a man possessed into further designs, even as he worked on a dozen other priority projects. Korolev set in place the infrastructure and the development of the critical path technologies for the Mars expedition: the isolation chamber, the *bochka*; the biospheres; the nuclear engines. All made progress: the *bochka* has been tested many times, though eclipsed by real spaceflights of long duration; the Bios programmes in Krasnoyarsk, matched by the real development of plants in orbital space stations; and the nuclear engines which were built and made long, successful test runs. Now the successor to Korolev's design bureau, Energiya, dedicated to his memory, has reclaimed this history and these critical paths. Energiya had updated the TMK, many times over: 1969 (MEK), 1986–1989, 1999, 2001 (MARSPOST) and 2005. Work on the critical paths has continued, like the trusses and structures on the Mir orbital station.

The unmanned Soviet programme for the exploration of the planets was not long under way when Sergei Korolev quickly ran into the problem that was to dog its development: unreliable upper stages. The early Soviet planetary programmes coincided with the early period of development of the 8K78, 8K78M and UR-500K rockets. Nowadays reliable, they suffered during their development phase from what would now be regarded as a shocking rate of malfunction. In the worst period, 1962, five out of six interplanetary probes were lost to rocket failure. Korolev undoubtedly acted with excessive haste, almost knowing that he had only a few years left to live and that he must work at a frenzied level to achieve anything at all. He would beat faults out of the system by sheer attrition. Haste is not the only explanation, for an important clue can be found in the recent memoirs of Boris Chertok. When he visited Germany in 1945–6, studying German achievements in science and technology, he was certainly struck by the German advances in rocket engines. But he was much more impressed by their level of development in tooling, manufacturing, precision mechanics, metalwork, guidance, gyroscopes and timers [1]. Here were skills, processes and traditions that had taken generations to build. Russia was far behind. These were the things that went wrong so many times with Korolev's 8K78 and 8K78M and Chelomei's Proton right up to 1972 – repeated, sadly, in 1996. Even then, the space probes themselves fell short of expectations, with communications failures at early stages (Venera 1) or late stages (Mars 1), equipment failures (Zond 2) and other breakdowns (Zond 1, Venera 2, 3). The scientific returns were meagre, only Mars 1 returning any significant data on interplanetary space.

The transfer of the interplanetary programme from Korolev's over-burdened design bureau to the Lavochkin design bureau was a major landmark, making possible a slower but more deliberate pace of development with an improved chance of success. The Brezhnev period ushered in an ultimately less capricious planning environment for space missions. Careful redesign and perseverance eventually paid off when Venera 4 entered the Venusian atmosphere in October 1967, followed by the Venera 5–6 twins in 1969. Now it was possible to obtain accurate, albeit disheartening and challenging scientific data on the conditions of Venus. The Venera probes were redesigned again and again so that they would eventually reach the surface and test the conditions there. These became the achievements of Venera 7 in 1970 and Venera 8 two years later, the latter achieving an impressive scientific return.

The Mars spacecraft of 1969–73, the first to be wholly indigenously engineered by the Lavochkin design bureau, are most remembered for their disappointments and difficulties. Although these certainly happened, their memory deserves better. By the standards of the early 21st century, they might appear primitive, but they were sophisticated spacecraft for their day, probably ahead of their time, carrying a large suite of scientific instruments for the study of Mars. The landers attempted to solve a problem no less challenging than the landing on Venus. Setting down a probe on Mars was doubly difficult, because of the imperfect knowledge of Mars's atmosphere and the precise trajectories necessary to ensure success. Despite that, Mars 2 became the first probe to impact on the planet and Mars 3 the first to make a soft-landing there. The orbiters executed a successful scientific programme in the nine months that followed, beaming back a broad range of data to Earth, an achievement reinforced by the mission of Mars 5 two years later. Contrary to the impressions of the time, there was a substantial scientific return from the Mars fleet of 1974: flyby pictures (Mars 4), a three-week orbital survey (Mars 5) and the first descent profile of the atmosphere (Mars 6).

The new Lavochkin Mars design came into its own when directed toward Venus. Now began more than ten years of success, the high summer of the Soviet interplanetary programme. Venera 9–14 made successful soft-landings on Venus, Venera 15 and 16 compiled radar maps while the VEGAs dropped more landers, balloons and went on to rendezvous with comet Halley in the most sophisticated mission ever undertaken in the inner solar system at the time. These spacecraft were advanced in their design, approach, equipment, techniques and systems used, especially those developed for the drilling, taking in and analyzing the soil. The scientific data collected were enormous, enabling scientists to profile the surface, rocks and atmosphere of the planet and, through the radar maps, begin to understand the larger forces at work shaping its evolution. Much of our knowledge of Venus to this day is directly derived from these Veneras.

The final Mars missions became, in the end, a by-word of the final days of the Soviet system and the difficulties of the new Russian Federation. Phobos 2 came tantalizingly close to success. Useful scientific information was returned and the manoeuvres undertaken to reach Phobos still impress. Even still, the Phobos and Mars 96 missions, so full of ambition and promise, were great disappointments and the programme ended, in the words of designer Vladimir Perminov, on a sour note.

For many, especially in the West, the overall impression of the Soviet and Russian interplanetary programme is a negative one, in which the failures are recalled with greater ease than the successes. At a time now when rockets have achieved high reliability rates (for some variants, 100%), it is easy to scoff at the repeated failures of the 8K78 upper stages at a time when rocket science was much less well understood. The successful landings on Mars in recent years look easy now, but they all benefitted from the difficult experiences of the early Mars and Venus probes. The great American planetary scientist Carl Sagan pointed out that both the United States and Soviet Union experienced high failure rates at the beginning. He looked at the record of the first 65 interplanetary and lunar launches of both countries. Both eventually achieved a cumulative launch success rate of 80%. The United States eventually achieved a mission success rate of 70% and Russia 60%, not as dramatic a difference between the two as some imagine [2].

The Russians were their own worst enemy, in some respects. We know now that they lost the race to the moon not because of technical incompetence but because of the unregulated rivalry of their design bureaus, poor organization, changing political priorities and allegiances, endless one-upmanship against the Americans, a chaotic institutional structure and the wasteful use of scarce resources. The Americans, by contrast, demonstrated disciplined planning, centralized organization (NASA) and the steady application of much greater resources in a cost-effective manner which gave them success not just in the moon programme but in the interplanetary missions of Mariner, Viking, Voyager, Pioneer Venus, *Magellan*, *Cassini* and the new Mars missions. In its interplanetary programme, the Russians displayed many of the faults that also characterized the moon programme. The scarce resources of the smaller Soviet economy were undoubtedly used wastefully, with the preparedness to continue multiple interplanetary launches in the face of similar, repeated failures. Until the victory of the 'Venusians' in the 'war of the worlds', many missions were organized on the basis of achieving specific objectives just ahead of the Americans, even if the technologies were not mature enough to match that ambition. Such planning, if it can be called that, was an over-riding feature of the Mars missions of 1969, 1971 and 1973, in the last extreme case sending spacecraft to the planet even when their transistors were known to be defective. Even after the disappointment of the Mars fleet, the

'Martians' were able to organize a counterattack, albeit with temporary success, but leading to the diversion of more wasted resources to the premature Mars sample return missions planned but wisely abandoned in the 1970s (5NM, 5M).

The scientific outcomes of Russian interplanetary missions were often under-estimated. In Europe and the United States, the discourse of popular science was largely one of the United States' voyage of discovery and science, a function of the American dominance of media, television, popular publishing and, it has to be said, an open space programme eager to publicize its achievements. Russia, by contrast, appeared to be running a secret programme that never published its scientific outcomes or results.

The reality, of course, was more complex and different. The Soviet space programme was well publicized in its mainstream promotional outlets, which ranged from Radio Moscow to booklets, magazines and other publications, in Russian, English and other languages. Scientific outcomes were published in in-house, Soviet and English language journals. On the Russian language side, the main outlet was a journal developed specifically for the purpose, *Kosmicheski Issledovania*, or *Cosmic Research*, and the principal outcomes of the main missions to Mars and Venus were published here in great detail. Sometimes an entire issue was devoted to the outcome of a particular set of missions. Soviet scientists published other articles in such places as *Proceedings of the Academy of Sciences*, *Geokhimiya*, *Analytische Khimiya*, and dedicated astronomy journals such as *Vestnik* and *Pisma*. The Nauka publishing house published books on the outcomes of Venus and Mars missions, such as *First panoramas of the surface of Venus* (1977) and *The planet Venus* (1989). Individual scientists published the outcomes of their experiments, or groups of them published collaborative analyses of a set of outcomes. Russian scientists published outcomes in a number of international, English language journals, such as the *Journal of Atmospheric Science*, *Icarus*, *Science*, *Nature* and *Planetary Space Science*. The weakness was that Russia was poor at promoting the outcomes of its missions in the non-scientific or popular media. There was no equivalent of, for example, the *National Geographic* magazine, with its superb photographs, commentary and ability to convey exploration to a non-specialized audience.

Although Russian cameras were of good quality, the same could not be said of the published versions of photographs, which were often reproduced in media outlets in quite degraded form. The first picture from the surface of Mars, taken in 1971, was not uncovered in the archives until 1999. The originals of the images taken of Venus by the Venera 9 and 10 orbiters have still not been seen. The *groza* 'sounds of Venus' were not released for acoustic interpretation until 2005. An image of Zond 1 was not released until 1996. As a result of all these factors, the scientific outcomes of Russian space missions were often underestimated. Western popular science writers did not learn Russian and as a result the published outcomes, especially in *Kosmicheski Issledovania*, went largely unnoticed. Not by NASA, though, which from its early days began what was called the *TTF Journal* – a series of translations of all the Russian scientific results. What circulation they achieved beyond NASA is not known.

What was missing in all this was a synthesis of the outcomes of the scientific endeavours of the two spacefaring nations. These two nations published in different places and apart from the limited range of English language journals, these rarely intersected. Once the Cold War was over, things began to change for the better. Collaborative articles began to appear between American and Russian scientists, putting together their respective experiences and discoveries. Several leading scientists from the old Soviet space programme took up academic posts in American universities. The University of Arizona Press published a series on the planets, one dedicated to each, but integrating the results of the missions of both countries. Perhaps the best example was in the area of Venus studies. Here, in 1981, Mikhail Marov had published the outcomes of Soviet studies of Venus, dedicated to the memory of his mentor and president of the Academy of Sciences, Mstislav Keldysh. When Mikhail Marov revised Soviet results in the exploration of Venus in the light of the missions in the 1980s, he did so in collaboration with an American writer and they brought together the results of not only the Venera missions, but also the American Pioneer Venus and *Magellan* missions [3].

With the fall of funding for the Russian space programme in the 1990s, even the protection of its scientific outcomes came into question. Ironically, it was NASA that came to the rescue. NASA began to compile a web archive of the results of Soviet lunar and interplanetary missions and put additional resources into data restoration. NASA asked for the old datasets from Moscow and quite a number began to arrive. These included, for example: particle counts, starting with Zond 1 in 1964 and the Mars, Venera and VEGA missions; proton counts from Zond 1 and 3; and energetic particle counts from Zond, Mars, Venera, VEGA and Phobos. The Zond 1 particle counts were made available to NASA by Moscow State University in 1999, some 35 years after the mission.

There are still many things we don't know about the Soviet interplanetary programme. Although the early Soviet space programme up to 1959 was quite open, a politically much more tightly controlled system of information flow curtailed our knowledge of the early missions. Some of the institutional memory of the programme seems to have been lost, one must fear forever, when construction of interplanetary spacecraft moved from Sergei Korolev's design bureau in April 1965 to the Lavochkin bureau. Even in later years, there are things we don't know. Photographs are not available of some of the early spacecraft, such as the 1M series in 1960. No information appears to be available on the later stages of the flights of some spacecraft (e.g., Zond 1–3). Explanations of the difficulties experienced on some missions are incomplete, imprecise or otherwise unconvincing (e.g., Venera 1–3). Our understanding of the manned programme benefitted substantially from the memoires of its leading personalities (e.g., chief designer Vasili Mishin, cosmonaut squad commander General Kamanin). On the interplanetary programme, two veterans have come forward – Roald Sagdeev and Vladimir Perminov – and their knowledge of particular events and programmes has been most helpful in filling definite gaps. But there are still wide-open spaces and many of the key protagonists are with us no more. Sergei Korolev is long dead (1966), while the key Lavochkin designers died at their posts (Babakin, 1971; Kovtunenko, 1993) or later (Kryukov, 2003) without having the time or opportunity to present their recollections.

The Soviet interplanetary programme does not only inform the planning of future missions to Mars – the design studies from the TMK onward are of continued relevance – but it is also important for the level of knowledge and science obtained. Venus and Mars are no longer as Tikhov left them in 1960: they are very different places, disappointing from the point of view of life and habitation, but of enduring fascination to planetary scientists. Such a changed knowledge is an undoubted achievement of – to a greater extent – the Soviet Venus and – to a lesser extent – Mars probes. The interplanetary programme was also an important part of the narrative of the overall Soviet space programme, even if it was eclipsed by events on a much grander scale in the moon programme and the human story of the courageous cosmonauts who flew around the Earth from 1961.

One day, in fulfilment of the dream of Tsiolkovsky, the pledge of Tsander and the schemes of Kondratyuk, Russian cosmonauts will travel to Mars. Whether they will do so with American astronauts or European or other colleagues, or after them, or whether they will meet Chinese *yuhangyuan* once they arrive, remains to be seen. But, once they do, the first part of the story of the adventure begun here will come to an end.

REFERENCES

[1] Chertok, Boris: *Rockets and people*, Vol. 1. NASA, Washington DC, 2005.
[2] Sagan, Dr Carl: *Pale blue dot*. Headline, London, 1995.
[3] Marov, Mikhail Y. and Grinspoon, David H.: *The planet Venus*. Yale University Press, New Haven, CT, 1998.

Appendix A

Soviet and Russian planetary missions

Grouped under launch windows

Related deep space Zond missions included

Date	Target	Designator	Outcome
10 Oct 1960	Mars flyby	Unannounced	Third-stage failure, reached 120 km
14 Oct 1960	Mars flyby	Unannounced	Similar
4 Feb 1961	Venus lander	*Tyzhuli sputnik*	Fourth-stage failure
12 Feb 1961	Venus lander	A.I.S./Venera 1	Contact lost, passed Venus 100,000 km
25 Aug 1962	Venus lander	Unannounced	Fourth-stage failure
1 Sep 1962	Venus lander	Unannounced	Fourth-stage failure
12 Sep 1962	Venus flyby	Unannounced	Third-stage explosion before orbit
24 Oct 1962	Mars flyby	Unannounced	Third-stage explosion before orbit
1 Nov 1962	Mars flyby	Mars 1	Passed Mars, May 1963
4 Nov 1962	Mars lander	Unannounced	Fourth-stage failure
11 Nov 1963	Technology test	Cosmos 21	Zond mission, fourth-stage failure
19 Feb 1964	Technology test	Unannounced	Zond mission, third-stage failure
27 Mar 1964	Venus flyby	Cosmos 27	Fourth-stage failure
1 Apr 1964	Venus lander	Zond 1	Passed Venus, contact lost
30 Nov 1964	Mars flyby	Zond 2	Contact lost
18 Jul 1965	Technology test	Zond 3	Passed moon on deep space trajectory
12 Nov 1965	Venus flyby	Venera 2	Passed Venus, contact lost
16 Nov 1965	Venus lander	Venera 3	Reached surface of Venus, contact lost
23 Nov 1965	Venus lander	Cosmos 96	Fourth-stage failure
(26 Nov 1965	Venus flyby	Unannounced	Unable to launch during window)
12 June 1967	Venus lander	Venera 4	Parachute descent (93 min)
17 June 1967	Venus lander	Cosmos 167	Fourth-stage failure
5 Jan 1969	Venus lander	Venera 5	Parachute descent (53 min)
10 Jan 1969	Venus lander	Venera 6	Parachute descent (51 min)

Date	Target	Designator	Outcome
27 Mar 1969	Mars orbiter	Unannounced	Second-stage failure
2 Apr 1969	Mars orbiter	Unannounced	First-stage failure
17 Aug 1970	Venus lander	Venera 7	Soft-landed, 23 min transmission
22 Aug 1970	Venus lander	Cosmos 359	Orbital failure
10 May 1971	Mars orbiter	Cosmos 419	Fourth-stage failure
19 May 1971	Lander/orbiter	Mars 2	Orbited Mars, landing failed
28 May 1971	Lander/orbiter	Mars 3	Orbited Mars, signals from surface
27 Mar 1972	Venus lander	Venera 8	Soft-landed, 63 min transmission
31 Mar 1972	Venus lander	Cosmos 482	Orbital failure
21 July 1973	Mars orbiter	Mars 4	Passed by at 1,300 km, pictures taken
25 July 1973	Mars orbiter	Mars 5	Orbited Mars successfully
5 Aug 1973	Mars lander	Mars 6	Lander made descent profile
9 Aug 1973	Mars lander	Mars 7	Failed to deploy lander, flew past
8 Jun 1975	Venus lander	Venera 9	Soft-landed, 56 min transmission
14 Jun 1975	Venus lander	Venera 10	Soft-landed, 66 min transmission
9 Sep 1978	Venus lander	Venera 11	Soft-landed, 95 min transmission
14 Sep 1978	Venus lander	Venera 12	Soft-landed, 110 min trasnsmission
30 Oct 1981	Venus lander	Venera 13	Soft-landed, transmitted 127 min
4 Nov 1981	Venus lander	Venera 14	Soft-landed, transmitted 57 min
2 June 1983	Venus orbiter	Venera 15	Radar mapper
7 June 1983	Venus orbiter	Venera 16	Radar mapper
15 Dec 1984	Venus lander	VEGA 1	Lander (56 min transmission), balloon
21 Dec 1984	Venus lander	VEGA 2	Lander (57 min transmission), balloon
7 Jul 1988	Mars orbiter	Phobos 1	Contact lost September
12 Jul 1988	Mars orbiter	Phobos 2	Intercepted Phobos, landing failed
16 Nov 1996	Orbiter/lander	Mars 8	Fourth-stage failure

Appendix B

Where are they now?

Table B.1. Landing coordinates of Soviet spacecraft on other worlds

Probe	Latitude	Longitude	Region
Venus			
Venera 3	−20°N to 20°N	60° to 80°E	(impacted)
Venera 4	19°N	38°	*Eisila* (destroyed during descent)
Venera 5	3°S	18°	*Navka Planitia* (destroyed during descent)
Venera 6	5°S	23°	*Navka Planitia* (destroyed during descent)
Venera 7	5°S	351°	*Navka Planitia*
Venera 8	10°S	335°	*Navka Planitia*
Venera 9	31.7°N	290.8°	*Beta Regio*
Venera 10	16°N	291°	*Beta Regio*
Venera 11	14°S	299°	*Navka Planitia*
Venera 12	7°S	294°	*Navka Planitia*
Venera 13	7°30′S	303°11′	*Navka Planitia*
Venera 14	13°15′S	310°09′	*Navka Planitia*
VEGA 1	7°11′N	177°48′	*Mermaid Plains*
VEGA 2	6°27′S	181°5′	*Aphrodite Mountains*
Mars			
Mars 2	44.2°S	213°W	*Eridania*
Mars 3	44.9°S	160.08°W	*Electris* and *Phaetonis*
Mars 6	23.9°S	19.4°W	*Mare Erythraeum*

Table B.2. Interplanetary missions orbiting the Sun

Venera 1
Mars 1
Zond 1
Zond 2
Zond 3
Venera 2
Mars 4
Mars 6
Mars 7 orbiter
Mars 7 lander
VEGA 1
VEGA 2
Phobos 1

Table B.3. Orbiting Venus

Venera 9
Venera 10
Venera 15
Venera 16

Table B.4. Orbiting Mars

Mars 2
Mars 3
Mars 5
Phobos 2
(Phobos 2 APS)

Table B.5. Crushed in Venusian atmosphere

VEGA 1 balloon
VEGA 2 balloon

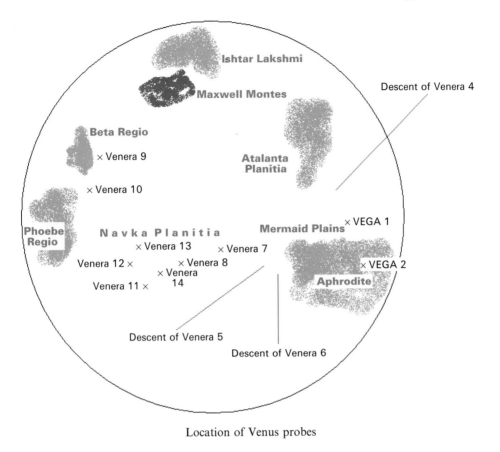

Location of Venus probes

Appendix C

Bibliography

BOOKS

Babakin, N.G., Banketov, A.N. and Smorkalov, V.N.: *G.N. Babakin, life and works*. Adamant, Moscow, 1996.

Burchitt, Wilfred and Purdy, Anthony: *Gagarin*. Panther, London, 1961.

Borisov, M.: *The craters of Babakin*. Znanie, Moscow, 1982.

Cattermole, Peter and Moore, Patrick: *Atlas of Venus*. Cambridge University Press, Cambridge, UK, 1997.

De Galiana, Thomas: *Concise Collins Encyclopaedia of Astronautics*. Collins, Glasgow, 1968.

Gatland, Kenneth: *Robot explorers*. Blandford, London, 1974.

Glushko, Valentin P:
  - *Development of rocketry and space technology in the USSR*. USSR Academy of Sciences, Novosti, Moscow, 1973.
  - *Rocket engines GDL-OKB*. USSR Academy of Sciences, Novosti, Moscow, 1979.

Harford, Jim: *Korolev*. John Wiley & Sons, New York, 1996.

Kieffer, H.H., Jakovsky, B.M., Snyder, C.W. and Matthews, M.S.: *Mars*. University of Arizona Press, Tucson, AZ, 1992.

Marov, Mikhail Y. and Grinspoon, David H.: *The planet Venus*. Yale University Press, New Haven, CT, 1998

Moore, Patrick: *On Mars*. Cassell, London, 1998.

Moore, Patrick: *The Guinness book of astronomy*, 5th edition. Guinness Publishing, Enfield, UK, 1995.

Perminov, Vladimir: *The difficult road to Mars – a brief history of Mars exploration in the Soviet Union*. Monographs in Aerospace History, No. 15. NASA, Washington DC, 1999.

Riabchikov, Yevgeni: *Russians in space*. Weidenfeld & Nicolson, London, 1972.

Sagdeev, Roald Z.: *The making of a Soviet scientist*. John Wiley & Sons, New York, 1994.

Semeonov, Yuri: *RKK Energiya dedicated to Sergei P. Korolev 1946–96*. RKK Energiya, Moscow, 1996.

Siddiqi, Assif: *The challenge to Apollo*. NASA, Washington DC, 2000.

Sidorenko, A.V. (ed.): *Poverkhnost Marsa*. Nauka, Moscow, 1980.

Stoiko, Michael: *Soviet rocketry – the first decade of achievement*. David & Charles, Newton Abbot, UK, 1970.

Surkov, Yuri: *Exploration of terrestrial planets from spacecraft – instrumentation, investigation, interpretation*, 2nd edition. Wiley/Praxis, Chichester, UK, 1997.

Turnill, Reginald: *Observer's book of unmanned spaceflight*. Frederick Warne, London, 1974.

USSR probes space. Novosti, Moscow, 1967.

Yumansky, S.P.: *Kosmonautika – Segondniya i zavtra*. Prosveshchenie, Moscow, 1986.

JOURNAL ARTICLES, ARTICLES, REPORTS, PAPERS, BROCHURES AND SIMILAR PUBLICATIONS

Balebanov, V.M., Zakharov, A.V., Kovtunenko, V.M., Kremev, R.S., Rogovsky, G.N., Sagdeev, R.Z. and Chugarinova, T.A.: *Phobos multi-disciplinary mission*. Academy of Sciences, Space Research Institute, Moscow, 1985.

Ball, Andrew: *Automatic interplanetary stations*. Paper presented to the British Interplanetary Society, 7th June 2003.

Ball, Andrew: *Phobos Grunt – an update*. Paper presented to the British Interplanetary Society, 5th June 2004.

Barsukov, V.L.: *Basic results of Venus studies by VEGA landers*. Institute of Space Research, Moscow, 1987.

Basilevsky, Alexander: The planet next door. *Sky and Telescope*, April 1989.

Beatty, J. Kelly: A radar tour of Venus. *Sky and Telescope*, May/June 1985.

Belitsky, Boris: How the soft landing on Mars was accomplished. *Soviet Weekly*, 15th January 1972.

Bond, Peter: *Mars and Phobos*. Paper presented to the British Interplanetary Society, 3rd June 1989

Breus, Tamara: *Venus – the only non-magnetic planet with a magnetic tail*. Institute for Space Research, Moscow, undated.

Burnham, Darren and Salmon, Andy:
 – Mars 96 – Russia's return to the red planet. *Spaceflight*, vol. 38, #8, August 1996.
 – On the long and winding road to Mars. *Spaceflight*, vol. 38, #11, November 1996.

Carrier, W. David III: *Soviet rover systems*. Paper presented at space programmes and technology conference, American Institute of Aeronautics and Astronautics, Huntsville, AL, 24th–26th March 1992. Lunar Geotechnical Institute, Lakeland, FL.

Central Intelligence Agency: *Soviet options for a manned Mars landing mission – an intelligence assessment*. Director of Intelligence, CIA, Washington DC, 1989.

Clark, Phillip S.:
 – Launch failures on the Soviet Union's space probe programme. *Spaceflight*, vol. 19, #7–8, July–August 1977.
 – The Soviet Mars programme. *Journal of the British Interplanetary Society*, vol. 39, #1, January 1986.
 – The Soviet Venera programme. *Journal of the British Interplanetary Society*, vol. 38, #2, February 1985 (referred to as Clark, 1985–6).
 – *Block D*. Paper presented to the British Interplanetary Society, 5th June 1999.

Congress of the United States: *Soviet space programs, 1976–80 – unmanned space activities*. 99th Congress, Washington DC, 1985.

Corneille, Philip: Mapping the planet Mars. *Spaceflight*, vol. 47, July 2005.

Covault, Craig: Soviets in Houston reveal new lunar, Mars, asteroid flights. *Aviation Week and Space Technology*, 1st April 1985.

Dollfus, A., Ksanformaliti, L.V. and Moroz, V.I.: Simultaneous polarimetry of Mars from Mars 5 spacecraft and ground-based telescopes, in M.J. Rycroft (ed.): *COSPAR Space Research*, papers, vol. XVII, 1976.

European Space Agency (ESA) & Centre National d'Etudes Spatiales: *VESTA – a mission to the small bodies of the solar system: report on the phase A study*. Paris, 1988.

Flight of the interplanetary automatic stations Venera 2 and Venera 3. *Pravda*, 6th March 1966, translated for NASA Goddard Space Flight Centre, 1966.

Friedman, Louis: *To Mars via Kamchatka*. Unpublished paper by the Planetary Society.

Goldman, Stuart: The legacy of Phobos 2. *Sky and Telescope*, February 1990.

Gordon, Yefim and Komissarov, Dmitry: *Illyshin-18, -20, -22 – a versatile turboprop transport*. Midland Counties, Hinckley, UK, 2004.

Gorin, Peter A.: Rising from a cradle – Soviet perceptions of spaceflight before Gagarin, in Roger Launius, John Logsdon and Robert Smith (eds): *Reconsidering Sputnik – forty years since the Soviet satellite*. Harwood, Amsterdam, 2000.

Gringauz, K.I., Bezrukih, V.V. and Mustatov, L.S.: Solar wind observations with the aid of the interplanetary station Venera 3. *Kosmicheski Issledovanya*, vol. 5, #2, translated by NASA Goddard Space Flight Centre, 1967.

Hansson, Anders: *V.I. Vernadsky, 1863–1945*. Paper presented to the British Interplanetary Society, 2nd June 1990.

Hansson, Anders: *The Mars environment in Russia*. Paper presented to the British Interplanetary Society, 12th June 1993.

Hansson, Anders: *Russian nuclear propulsion*. Paper presented to the British Interplanetary Society, 25th May 2002.

Huntress, W.T., Moroz, V.I. and Shevalev, I.L.: Lunar and robotic exploration missions in the 20th century. *Space Science Review*, vol. 107, 2003.

IKI (Institute of Space Research): *The Soviet programme of space exploration for the period ending in the year 2000 – plans, projects and international cooperation. Part 2: The planets and small planets of the solar system*. Institute of Space Research, USSR Academy of Sciences, 1987.

Illyin, Stanislav: From project VEGA to project Phobos. Novosti Press Agency Soviet Science and Technology *Almanac*, 1987.

Ivanovsky, Oleg: Memoir, in John Rhea (ed.): *Roads to space – an oral history of the Soviet space programme*. McGraw-Hill, London, 1995.

Johnson, Nicholas L.: Soviet atmospheric and surface Venus probes. *Spaceflight*, vol. 20, #6, June 1978.

Kemurdzhian, A.L., Gromov, V.V., Kazhakalo, I.F., Kozlov, G.V., Komissarov, V.I., Korepanov, G.N., Martinov, B.N., Malenkov, V.I., Mityskevich, K.V., Mishkinyuk, V.K. *et al.*: Soviet developments of planet rovers 1964–1990. CNES and Editions Cepadues: *Missions, technologies and design of planetary mobile vehicles*, proceedings of conference, Toulouse, September 1992.

Kerzhanovich, Viktor V.: Improved analysis of descent module measurements. *Icarus*, vol. 30.

Khrushchev, Sergei: The first Earth satellite – a retrospective view from the future, in Roger Launius, John Logsdon and Robert Smith (eds): *Reconsidering Sputnik – forty years since the Soviet satellite*. Harwood, Amsterdam, 2000.

Kidger, Neville:
  – Phobos mission ends in failure. *Zenit*, #27, May 1989.
  – Phobos update. *Zenit*, #25, March 1989.
  – Project Phobos – a bold Soviet mission. *Spaceflight*, vol. 30, #7, July 1988.
Klaes, Larry: Soviet planetary exploration. *Spaceflight*, vol. 32, #8, August 1990.
Kondratyev, K.Ya. and Bunakova, A.M.: *The meteorology of Mars*. Hydrometeorological Press, Leningrad, 1973, as translated by NASA, TTF 816.
Kopik, A.: Big plans at NPO Lavochkin. *Novosti Kosmonautiki*, vol 15, #10, 2005.
Kotelnikov, V.A., Petrov, B.N. and Tikhonov, A.N.: Top man in the theory of cosmonautics. *Science in the USSR*, #1, 1981.
Kovtunenko, Vyacheslav M., Kremev, Roald S., Rogovsky, G.N. and Sukhanov, K.G.: *Combined programme of Mars exploration using automatic spacecraft*. Babakin Research Centre, Moscow, 1987.
Kovtunenko, Vyacheslav M.: *Achievements of science and engineering of the USSR in the exploration of the planet Venus by the use of spacecraft*. Intercosmos Council, Moscow, 1985.
Kovtunenko, V., Kremev, R., Rogovsky, G. and Ivshchenko, Y.: Prospects for using mobile vehicles in missions to Mars and other planets. Babakin Research Centre, Moscow, published by CNES and Editions Cepadues: *Missions, technologies and design of planetary mobile vehicles*, proceedings of conference, Toulouse, September 1992.
Kulikov, Stanislav: Top priority space projects. *Aerospace Journal*, November–December 1996.
Kuzmin, Ruslan and Skrypnik, Gerard: A unique map of Venus. Novosti Press Agency Soviet Science and Technology *Almanac*, 1987.
Langereux, Pierre: Les quatre sondes Franco-Soviétiques VESTA vont explorer Mars et huit petits corps. *Air et Cosmos*, #1117, novembre 1986.
Lardier, Christian: Revision des future programmes martiens. *Air et Cosmos*, vol 1518, 12 mai 1995.
Lardier, Christian: CADB devoile le moteur nucleare RD-0410. *Air et Cosmos*, vol. 1571, 21 juin 1996.
Lardier, Christian: Les nouveaux projets de la NPO Lavochkine. *Air et Cosmos*, vol. 1609, 18 avril 1997.
Lardier, Christian: Le Jimo, version russe. *Air et Cosmos*, vol. 1955, 22 octobre 2004.
Lenorovitz, Jeffrey M.: France designing spacecraft for Soviet interplanetary missions. *Aviation Week amd Space Technology*, 7th October 1985.
Lepage, Andrew L.: The mystery of Zond 2. *Journal of the British Interplanetary Society*, vol. 46, #10, October 1993.
Les moteurs nucléaires de l'ère soviétique. *Air et Cosmos*, vol. 1810, 21 septembre 2001.
Lovell, Bernard:
  – *The story of Jodrell Bank*. Oxford University Press, London, 1968;
  – *Out of the zenith – Jodrell Bank, 1957–70*. Oxford University Press, London, 1973.
Mars 96 sera le dernier poids lourd martien. *Air et Cosmos*, vol. 1587, 15 novembre 1996.
Moroz, V.I.: The atmosphere of Mars. *Space Science Review*, vol. 19, 1976.
Moroz, V.I., Huntress, W.T. and Shevalev, I.L.: Planetary missions of the 20th century. *Cosmic Research*, vol. 40, #5, 2002.
Oberg, Jim: The probe that fell to Earth. *New Scientist*, 6th March 1999.
On course to meet the comet. *Soviet Weekly*, 21st December 1985.
One day we shall fly to Mars. *Soviet Weekly*, 27th August 1977.
Phelan, Dominic: Russian space medicine still aims for Mars. *Spaceflight*, vol. 46, #1, January 2004.

Poliakov, Dr Valeri: Remarks made at presentation in British Interplanetary Society, London, 22nd May 2002.

Popov, G.A., Obukhov, V.A., Kulikov, S.D., Goroshkov, I.N. and Upensky, G.R.: *State of the art for the Phobos Soil return mission.* Paper presented to 54th International Astronautical Congress, Bremen, Germany, 29th September–3rd October 2003.

Prismakov, Vladimir, Abramorsky, Yevgeni and Kavelin, Sergei: *Vyacheslav Kovtunenko – his life and place in the history of astronautics.* American Astronautical Society, history series, vol. 26, 1997.

Rachuk, Vladimir: Best rocket engines from Voronezh. *Aerospace Journal*, November/December 1996.

Results of the Phobos project. Soviet Science and Technology *Almanac* 90. Novosti, Moscow, 1990.

Robertson, Donald F.: Venus – a prime soviet objective. *Spaceflight*, vol. 34, #5, May 1992.

Sagdeev, Roald: Halley's comet – the VEGA story. *Spaceflight*, vol. 28, #11, November 1986.

Salmon, Andy: *Mars 96 – the Martian invasion.* Paper presented to the British Interplanetary Society, 1st June 1996.

Salmon, Andy and Ball, Andrew: *The OKB-1 planetary missions.* Paper presented to the British Interplanetary Society, 2nd June 2001.

Science and space. Novosti, Moscow, 1985.

Serebrennikov, V.A., Stekolshikov, Y.G., Iljin, M.N. and Shevaliov, I.L.: *Design concepts and utilization of the propulsion system of the spacecraft in the Phobos project.* Institute of Space Research, Moscow, 1988.

Siddiqi, Asif: *Deep space chronicle.* NASA, Washington DC, 2001.

Siddiqi, Asif: A secret uncovered – the Soviet decision to land cosmonauts on the moon. *Spaceflight*, vol. 46, #5, May 2004.

Smid, Henk: Soviet space command and control. *Journal of the British Interplanetary Society*, vol. 44, #11, November 1991.

Soviet space odyssey. *Sky and Telescope*, October 1985.

Telegin, Y.: Preparing for the Mars 94 mission. *Spaceflight*, vol. 35, #9, September 1993.

Tyulin, Georgi: Memoirs, in John Rhea (ed.): *Roads to space – an oral history of the Soviet space programme.* McGraw-Hill, London, 1995.

Varfolomeyev, Timothy:
  – The Soviet Venus programme. *Spaceflight*, vol. 35, #2, February 1993.
  – The Soviet Mars programme. *Spaceflight*, vol. 35, #7, July 1993
  – Soviet rocketry that conquered space. *Spaceflight*, in 13 parts:
    1 Vol. 37, # 8 August 1995;
    2 Vol. 38, #2, February 1996;
    3 Vol. 38, #6, June 1996;
    4 Vol. 40, #1, January 1998;
    5 Vol. 40, #3, March 1998;
    6 Vol. 40, #5, May 1998;
    7 Vol. 40, #9, September 1998;
    8 Vol. 40, #12, December 1998;
    9 Vol. 41, #5, May 1999;
    10 Vol. 42, #4, April 2000;
    11 Vol. 42, #10 October 2000;
    12 Vol. 43, #1, January 2001;
    13 Vol. 43, #4 April 2001 (referred to as Varfolomeyev, 1995–2001).

Veha reveals Venus' secrets. *Soviet Weekly*, 7th December 1985.

Venus – 470C in the Sun! *Soviet Weekly*, 16th September 1972.

Zaitsev, Yuri: Preparing for Mars – a simulated manned mission to the red planet is about to begin. *Spaceflight*, vol. 47, #1, January 2005.

Zaitsev, Yuri: The successes of Phobos 2. *Spaceflight*, vol. 31, #11, November 1989.

Zygielbaum, Joseph L. (ed.): *Destination Venus – communiqués and papers from the Soviet press, 12th February to 3rd March 1961*. Astronautics Information, Translation 20. Jet Propulsion Laboratory, Pasadena California.

BOOKLETS

Road to the red planet. Novosti, Moscow, 1988.
Road to the stars. Novosti, Moscow, 1986.
Science and space. Novosti, Moscow, 1985.
Soviet space research. Novosti, Moscow, 1970.
Soviet space studies. Novosti, Moscow, 1983.
USSR probes space. Novosti, Moscow, 1967.

INTERNET

Darling, David: Gavril Tikhov at *http://www.daviddarling.info.encyclopedia/t/tikov*

Grahn, Sven: *www.svengrahn.ppe.se*
  – Jodrell Bank's role in early space tracking activities;
  – Soviet/Russian OKIK ground station sites;
  – The Soviet/Russian deep space network;
  – Yevpatoria – as the US saw it in the 60s;
  – A Soviet Venus probe fails – and I stumble across it.

Hendrickx, Bart: *Soviet Venus lander revealed*, Friends and Partners in Space posting, 30th August 2001.

Mitchell, Don P. (2003–4):
  – Soviet interplanetary propulsion systems;
  – Inventing the interplanetary probe;
  – Soviet space cameras;
  – Soviet telemetry systems;
  – Remote scientific sensors;
  – Biographies;
  – Plumbing the atmosphere of Venus;
  – Drilling into the surface of Venus;
  – Radio science and Venus;
  – The Venus Halley missions at *http://www.mentallandscape.com*

NASA: Mars 5, NASA NSSDC master catalogue *http://nssdc.gsfc.nasa.gov/database*. NASA, Washington DC, 2005.

Oberg, Jim: Are women up to the job of exploring Mars? *MSNBC*, 11th February 2005.

Present Russian Mars plans: *http://www.energia.ru.english/Energiya/mars*

Zak, Anatoli: Martian expedition, *http://www.russianspaceweb.com*

PERIODICALS AND JOURNALS CONSULTED

Air et Cosmos.
Aviation Week and Space Technology.
Ciel et Espace.
Flight International.
Icarus.
Soviet Weekly.
Space Science Review.
Spaceflight.

Index